T0146292

Glassfibre Reinforced Concrete

Glassfibre Reinforced Concrete
Principles, production, properties and applications

Professor Peter J. M. Bartos

WHITTLES PUBLISHING

Published by
Whittles Publishing,
Dunbeath,
Caithness KW6 6EG,
Scotland, UK

www.whittlespublishing.com

© 2017 Peter J. M. Bartos

ISBN 978-184995-326-9

Typeset by
Deanta Global Publishing Services, Chennai, India

Print managed by Jellyfish Solutions Ltd

Contents

List of figures

List of tables

Acknowledgements

The author wishes to thank all who helped him in the development of this publication, particularly Graham T. Gilbert for valuable information and comments, Peter Ridd for providing numerous technical and project data, Glyn Jones for input on structural design, Ian White for advice on production processes and for reading of the manuscript and Neil Sparrow for access to the International Glassfibre Concrete Association (iGRCA) pictorial archive.

The author wishes to acknowledge the inspiration and personal guidance provided by Professor Howard G. Allen of the University of Southampton at the very beginning of the development of GRC, whose predictions of wide use of GRC in the future have been now matched and even exceeded.

The book was produced with the assistance of the iGRCA, for which the author is grateful. However views expressed in this publication are those of the author and do not necessarily reflect those of the iGRCA.

Preface

The development of glassfibre reinforced concrete (GRC) in the 1960s [1–4] exploited an ancient and simple principle of converting naturally brittle materials into much tougher and therefore more useful ones through the incorporation of strong fibres, initially of plant origin. GRC is based on the same principle, and in the twenty-first century it is an already well-established construction material used all over the world.

Basic constituents of GRC are very few, namely cement, water, fine aggregate and glass fibres. However, the internal structure of the composite itself is as complex as that of the most advanced high-tech materials, such as composites used in the aerospace industry. Paradoxically, a high-performance material such as GRC can also be reliably produced using relatively simple and inexpensive processes.

The range of its applications is already very wide. In its basic form, it is used to produce simple items such as ornamental flowerpots; while in its high-tech version, it is the preferred construction material for the production of large, thin-walled structural elements of very complex shapes.

GRC is a cement-based composite strongly related to concrete. However, in order to exploit its outstanding properties, substantial additional knowledge and understanding are required. Applications which utilise high-performance GRC do not require an excessively sophisticated and expensive production plant, but they do require a very strict production regime. An adequate level of supervision is much closer to that used for the production of high-performance, polymer-based, fibre-reinforced composites than to the manufacture of ordinary precast reinforced concrete products.

GRC continues to develop (for example, eGRC). Numerous scientific and technical papers which focus on specific aspects of GRC exist. Many of these have been published together in the proceedings of congresses held by the

International Glassfibre Reinforced Concrete Association (iGRCA) every 2–3 years from 1977 [5] to date, while others have appeared in scientific and technical journals. However, all of the books which surveyed existing knowledge [6–8] were published in the earlier stages of GRC's development, the last one being published in the early 1990s.

Growth in the use of GRC since the turn of the twenty-first century has been and continues to be very strong. There is therefore a need to review the current stage of development and the applications of GRC. This book reviews historical background, indicates raw materials and outlines the different production processes and properties which can be achieved. Recent developments are highlighted, and the book illustrates the very wide range of GRC applications. At the same time, recent growth in the range and volume of practical applications has not been matched by advances in the understanding of this complex composite. Fundamental and unique aspects of the microstructure and fracture mechanism of GRC are outlined and discussed, without going into details, which are available in previous publications. This book shows that the full potential of GRC as a structural material has not yet been realised, and that a comprehensive understanding of GRC has yet to be achieved. An improved understanding is a prerequisite for successfully taking on the challenge of further improving its already outstanding properties and providing a solid background to an even greater use of advanced GRC in practical construction.

It is a big challenge, requiring not only the highest level of investigative skills and background knowledge on the part of the researcher, but expensive, state-of-the-art research facilities as well. Compared to other widely used construction materials, GRC is one which has so far benefited the most from the exploitation of nanotechnology such as admixtures designed at molecular scale, investigation of bond using nanoscale apparatus, and nanoparticles in photocatalytic surfaces.

Most importantly, substantial additional funding is required to make all the advances achieved in basic research useable by practitioners in the construction industry. Unfortunately, design and manufacture of GRC are carried out almost entirely by small- and medium-sized companies, which even in the best of times do not have adequate internal resources to support research and development (R&D). It is therefore essential that national and international R&D funding authorities recognise this and provide assistance to the GRC industry to keep it moving forward.

PJMB

Glossary

Additive
Material other than cement or aggregate added to the GRC mix in a significant proportion (usually >5% by weight).

Admixture
A substance added in small quantities, usually <4% by weight of cement, to the fresh mix in order to modify the properties of the GRC in fresh or hardened state.

Aggregate/cement ration
The ratio of the mass of total saturated surface dry aggregate to the mass of dry cement.

Alkali-resistant glass (AR glass) fibre
Fibre drawn from glass containing a minimum of 16% b.w. of zirconium dioxide.

AMS – Approved Manufacturer Scheme
A system of accreditation of a GRC manufacturer's ability to provide resources, skills, and equipment necessary to meet the levels of quality required for GRC products. Replaced by the Full Member category of iGRCA membership from late 2016.

Anchor
A device for attachment of GRC skin to a supporting frame (stud frame system). There are gravity, flex and seismic anchors.

AR glass fibre
See Alkali-resistant glass fibre.

Backing mix
The bulk of the GRC sprayed into a mould after the initial mist coat, face mix and so on have been placed.

Bag and bucket tests
Methods for calibration of equipment for rate of delivery of fibres and matrix in production of GRC.

Bonding pad
An additional thickness of the GRC required to safely embed an anchor in the GRC skin.

BOP (value)
Stress at the 'bend-over-point' in a stress-strain or load-deformation curve recorded in a test. Used mainly in the USA.

b.w. and b.v.
Abbreviation of the expressions 'by weight' or 'by volume' when proportions or content of materials making up the GRC mix are stated. It is essential to understand the difference.

CCV
composite ciment verre = GRC in French.

Characteristic value
Value of a property which is expected to be exceeded by a proportion (taken as 95%, unless a different value is specified) of the population of all measurements of that property showing a normal distribution.

Compaction
Reduction of voids in a fresh, uncured composite to a practical minimum, usually using vibration, rolling, tamping, pressure or a combination of these methods.

Coupons
Specimens taken from a test board for the purpose of determining their mechanical properties.

Curing

A period of time required for a freshly made GRC element to remain in an adequately moist and warm environment in order to continue the hydration of cement and to develop minimum mechanical properties required.

Dry curing

A GRC mix containing an addition of an acrylic polymer (minimum 5% polymer solids by weight of cement) that dispenses with the need to place the GRC product in a wet/humid environment or seal it externally for up to 7 days.

Dry density

Mass/unit volume of the composite in an oven-dried condition.

E-glass fibre

Fibre based on common borosilicate glass, routinely used for reinforcement of polymers. Fibre is susceptible to alkali-related corrosion when embedded in an OPC-based matrix. It is not used for production of GRC.

*e*GRC

Glass-reinforced concrete with a photocatalytic surface. The active surface de-pollutes the surrounding air and provides the GRC with a high degree of self-cleaning capability in the presence of daylight (or another source of UV light).

Engineer

The person of authority responsible for the design of a GRC product.

Facing coat

An initial layer of a cementitious slurry sprayed into a mould. It may include decorative aggregate and/or pigments.

Fibre (Fiber in USA)

An individual filament of glass, usually with a diameter between 9 to 20 μm.

Fibre content by volume (% vf) or by weight (% wf)

The ratio of the volume (v) or mass (weight w) of fibres to the total unit volume/mass of GRC in an uncured state, expressed as a percentage.

Flex anchor

A flexible metal element embedded firmly in the GRC and fixed to a supporting frame/structure.

Full Member (of iGRCA)
A category of iGRCA membership from late 2016, which indicates the ability of a GRC manufacturer to provide resources, skills and equipment necessary to meet the levels of quality required for GRC products.

GFB
glasfaserbeton = GRC in German.

Ghosting
Effect of a variable thickness or different levels of compaction of GRC on the appearance of its external surface, visible namely when wet.

GRC (GFRC in USA)
An acronym for Glassfibre Reinforced Concrete (or Cement) – a composite material produced by reinforcement of a cementitious matrix with compatible glass fibres.

GRCA
see iGRCA.

Gravity anchor
Fixing which transfers the dead weight of a GRC element to its support (frame etc.).

High-shear mixer
Mixer with tools rotating at high speeds which generate high shearing action, that breaks up any agglomerations of cement and other particles and produces the highly workable cementitious slurry for manufacture of GRC by the spray-up process.

iGRCA
International Glassfibre Concrete Association, founded in October 1975.

Insert
A metallic connector with an internal thread embedded in GRC.

Limit of proportionality – LOP
The point at which the plot of stress versus strain begins to deviate from linearity, also known as the elastic limit and BOP in the USA.

Matrix
A cementitious slurry mixed with fine aggregate. One of the two basic components of GRC.

Metakaolin

A dehydroxylated form of the clay mineral kaolinite. The particle size is smaller than that of cement, but not as fine as microsilica. It acts as a reactive aluminosilicate pozzolan.

Minimum film formation temperature – MFFT

Temperature required for the acrylic polymer within the composite to enable it to form a very thin layer retaining moisture in the fresh GRC, permitting a 'dry curing' process.

Mist coat

A thin layer of cementitious slurry without glass fibres sprayed first into a mould to provide an initial coating, and then the surface layer of the GRC product.

Modulus of rupture – MOR

It is the highest stress recorded during a flexural (bending) test on a specimen of GRC. It is equal to flexural (bending) strength of the composite.

Polymer-modified GRC

GRC containing an acrylic thermoplastic polymer at min. 5% of polymer solids b.w. of cement.

Premix GRC

GRC produced by blending together pre-cut glass fibres and fresh cementitious slurry in a mixer.

Premix mixer

A two-stage mixer in which the cementitious slurry is produced first, followed by the blending-in of pre-cut glass fibres.

Producer

The person of authority entering into a contract to manufacture a GRC product.

Purchaser

The person of authority entering into a contract to buy a GRC product.

Retarder

An admixture which extends the setting time of fresh matrix mix.

Roving
A number of strands of glass fibres loosely gathered together and wound to form a cylindrical package.

Sand/cement ratio
The ratio of the mass of dry sand to cement in the GRC mix.

Sandwich panel
A panel with outer layers made of GRC, separated by a layer of insulating material.

Scrim
A lightweight fabric (net) made from widely spaced continuous strands laid over each other at ninety-degrees with a binder holding the fabric structure in place.

Sealant
An elastic material used to fill joints between GRC elements and prevent passage of water.

Self-compacting premix GRC
A fresh GRC mix with very high workability achieved by using special plasticisers. The mix can be poured and compacted into the moulds without the need of external mechanical vibration.

Serviceability limit state
A limit to the condition of a GRC structural element (usually a maximum allowable deflection) when subject to loads in service.

Size
A coating applied to individual glass fibres/filaments during their manufacture.

Slump test (mini-slump test)
A test for measurement of consistency of fresh slurry/cement paste to be used specifically for production of GRC. Not to be confused with the slump test for ordinary concrete.

Slurry
A freshly mixed matrix containing cement, sand, water, admixture and polymer (optional) ready for spraying or for fibres to be mixed in.

Specifier
The person of authority responsible for the specification of a GRC product.

Spray-up process
A method using a simultaneous spraying of fresh cementitious slurry and chopped glass fibres to produce GRC.

Sprayed premix process
A method in which glass fibres chopped at the spray head are added through the same spray head to produce GRC.

Strand
A bundle of individual filaments held together by their coating (size). Numbers of filaments in one strand can vary from approximately 50 to several hundred, the usual content being about 200.

Strand-in-cement (SIC) test
A test for durability of glass fibres embedded in a cement-based matrix.

Stud frame
A structural GRC cladding system in which panels made of a single sheet of GRC are attached to a metal supporting frame.

Superplasticiser
Admixture enabling a significant reduction of water content while workability is maintained and strength increased or for a significant increase in workability for the same water content.

Supplier
The person of authority entering into a contract to supply goods to the producer.

Support anchor
See Gravity anchor.

Test board
A sheet of GRC manufactured during a normal production run for the purpose of assessment of the quality of a product.

Test board mean
The arithmetic mean calculated from results of tests on all coupons from a single test board.

Test sample
The total number of coupons taken from a test board and tested to determine a property of GRC represented by that board.

Tex
Weight in grams of a 1000 m length of a fibre strand or of a roving.

Textile-reinforced concrete – TRC
A cementitious matrix reinforced with woven fibres (textiles) which may or may not be alkali-resistant.

Tolerance
A permissible variation between specified and actual value of a property, namely size of an element.

Top/bottom ratio
The ratio of the value of MOR obtained with the tensile face of the test specimen being the mould face to that of the back face. The 'tensile face' of the test specimen is the bottom one when in the machine for a bending /flexural test. The 'mould face' is that in contact with the mould (GRC is mostly made in sheet-like elements), it can be entirely smooth, or showing the pattern of the mould. The 'back face', is the opposite one, usually of a rougher surface after compaction by profiled rollers.

Trowelled surface
Surface of a GRC panel away from the mould face, often smoothed using a trowel.

Uncured state
Period in which the GRC composite has been fully formed, but the fibre can be still separated from the matrix by a stream of water.

Uniaxial tensile strength – UTS
Maximum stress obtained during a uniaxial tensile test.

Wash-out test
A test to determine the fibre content of fresh GRC.

Water/cement ratio
The ratio of the mass of total free water to the mass of cement in the GRC mix.

Water/binder ratio
The ratio of the total mass of free water to the total mass of cement and other binders (e.g. pozzolanic additions) in the GRC mix.

Introduction and Scope | 1

Glassfibre reinforced concrete (GRC) is arguably the most complex material widely used in current construction practice. It is an unusual composite, in which both the matrix and the reinforcement alone can be themselves considered composites. The complexity of plain concrete is well appreciated, and its reinforcement by steel bars, introduced over a century ago, is well understood. However, GRC differs significantly from other cement-based composites such as the ordinary reinforced concrete mentioned above. Principal differences lie with the type of the reinforcement, nature of the internal fibre-matrix stress transfer and types of failure mechanism.

Reinforcement of the cementitious matrix in GRC is not generally achieved by dispersion or an organised positioning of single solid elements such as steel bars/rods or single fibres. Instead, the basic reinforcement consists of bundles of varying length and shape, containing different numbers of individual fibres/filaments [9]. The bundles are dispersed throughout the GRC, often in a very non-uniform manner in comparison with the precise positioning of reinforcement in traditional reinforced concrete.

Bond between reinforcement and matrix, the key requirement for the generation of a 'composite action' between them, governs both the properties of the reinforcing fibre bundles (bond between individual fibres/filaments in a bundle) and those of the GRC in bulk. Bond also always changes with time and according to service conditions in a non-uniform manner, as does the strength of the fibres and of the fibre bundles. GRC is therefore a very complex, genuinely 'high-tech' composite.

First attempts at production of 'modern' GRC were reported from the former Soviet Union [1], but the bulk of the initial development of GRC, which made it into a marketable product, took place in the UK. The early composite became known as glassfibre reinforced *cement*. Terminology changed gradually, and since the late 1990s the same material has been called glassfibre reinforced *concrete*, on account of the substantial content of fine aggregate now used in the mix. It is still abbreviated as GRC (or its translation) in most parts of the world. The only exception is North America and countries where US-construction terminology and spelling was adopted. There GRC is referred to as glassfiber reinforced concrete, abbreviated GFRC. Both GRC and GFRC apply to the very same material and the term is fully interchangeable. GRC is the term more widely used worldwide, and it has been therefore adopted throughout this publication.

Applications of GRC envisaged in early stages of its development were largely nonstructural. However, once its behaviour and fracture mechanisms became better understood, the range of applications widened to include 'semi-structural' (resisting self-weight alone) and eventually fully structural (resisting imposed loads) products. The latter is now supported by a structural design system reflecting the properties and fracture mechanisms of GRC.

This publication aims to go beyond the condensed technical information provided in technical literature from GRC material suppliers or manufacturers, mainly for marketing purposes. More detailed technical information is provided, explaining the unique aspects of internal structure and fracture mechanisms of GRC with references provided for those who wish to know more, namely when engaged in or planning research concerning GRC. There is potential for substantial improvement of properties of the GRC beyond those used in contemporary construction practice.

Concrete reinforced principally with a glassfibre fabric or textile, known as textile-reinforced concrete (TRC) differs significantly from GRC. The principal reinforcement in TRC is glass fibre strands acting as very long, continuous rather than short, discontinuous reinforcing elements, interwoven as a fabric/textile. The position of reinforcement in TRC is predetermined by the nature of its 'textile' pattern. Both the fracture mechanisms and processes used for production of TRC therefore differ from those of GRC. Existing GRC manufacturers are rarely involved in the production of pure TRC. At present, practical production of TRC is limited to a few specialist companies, mainly located in Germany. Glass fibre fabrics are sometimes used as an additional reinforcement in conjunction with short dispersed fibres, namely in a premix type of GRC production. However, concrete reinforced principally with glass fibres in a textile/fabric form (TRC) is outside of the scope of this document. Wider information can be obtained from publications focused specifically on TRC [e.g. 10, 11].

Alkali-resistant glass fibres can be used together with metallic fibres, usually as the secondary (and rarely as primary) reinforcement in concrete slabs serving as floors or pavements. Such concrete is classed as GRC if its reinforcement is purely by glass fibres. In other cases, it is classed as a type of fibre-reinforced concrete (FRC). Fibre concrete – FRC – usually denotes mixes where the prevailing fibrous reinforcement is metallic (e.g. steel) or polymeric fibres.

Brief History of Development | 2

First practical applications of the principle of making a naturally occurring brittle material less brittle and more ductile through the incorporation of strong, stiff fibres can be traced back for several millennia when hard, dry clay was reinforced with straw or other natural fibres. Fibre-reinforced brittle-matrix composites for use in construction are therefore nothing new. However, it was not until the invention of asbestos cement in the late nineteenth century that a 'modern' mass-produced, high-performance, cement-based, fibre-reinforced composite entered general construction practice.

The beginning of the twentieth century brought about technological advances in the production of glass fibres, mainly of the borosilicate type (E-glass), and the first polymers entered industrial production. Thermosetting polyester resins provided the matrix material for the production of glassfibre-reinforced polymer ('plastic'), abbreviated as 'GRP' and also known as 'fibreglass'.

A major expansion in the availability of different types of fibres after the Second World War encouraged development of new fibre-reinforced composites, which included those with cementitious matrices. The late 1950s brought about the first production of GRC in the USSR, albeit based on E-glass fibres combined with non-alkaline matrices [1]. Ordinary Portland cement could not be used as the matrix material, as its alkaline environment caused a serious corrosion and loss in tensile strength of the E-glass fibres; as a result the earliest GRC production remained very limited.

It had already been known at the time of the first GRC that special glass formulations which imparted different degrees of alkali resistance did exist. However, it was not until the late 1960s that such fibres became commercially available. A joint research project with a team led by A.J. Majumdar at the UK Building Research Station and A. Tallentire at Pilkington Glass Co., a major UK glassmaking company, were behind the development of commercially produced, alkali-resistant (AR) fibres, marketed as the Cem-Fil® fibre [4]. At the same time, Japanese glass manufacturers, namely the Nippon Electric Glass Co. (NEG) and Asahi Glass Co. (AGC), also developed the commercial production of AR fibres with a high content of ZrO_2. The new fibres were covered by patents and practical applications in early years and subject to licensing. The aim was to provide guidance on production and to ensure adequate quality of GRC when nobody had any practical experience in GRC technology.

Popularity of the new composite grew rapidly, and the Glassfibre Reinforced Cement Association (now the International GRCA) was established in 1975 as a global forum for spreading the know-how, exchanging experience and new information, and promoting GRC as a material offering significant new benefits to construction industry and to end users.

The rapid rise in acceptance of GRC was temporarily curtailed when problems with its performance emerged in early practical applications. In general, these were a direct reflection of a lack of quality control in the very early years of GRC production, which was much more sophisticated than expected. Manufacture was to be controlled by restricting it to holders of a licence from the fibre producer. However, the licensees were often small construction companies or new start-ups, inexperienced in dealing with a fundamentally high-tech material requiring significantly greater precision during production than that needed for ordinary reinforced concrete. Quality control rules attached to their licences were often inadequately observed and rarely checked. The situation was not helped by early versions of GRC showing a loss of the initial pseudo-ductility with age (embrittlement) when exposed to a wet environment.

Manufacturers of the AR glass fibres responded in the early 1980s by developing a new generation of fibres with different surface treatments. New fibres significantly improved the long-term performance of hardened GRC, especially when aged in a humid service environment. An example of this development was the introduction of CemFil2® AR fibres. Another alternative was to replace 10% or more of the cement in the GRC matrix with an acrylic polymer. This generated a more durable composite [12, 13], but the extra cost and reduced fire resistance greatly restricted its adoption in construction practice.

The early years of GRC's development coincided with the period of demise of traditional asbestos cement due to the carcinogenic nature of most of the asbestos fibres it used, combined with the brittle behaviour of aged products (e.g. corrugated roofing). An urgent need arose to develop an alternative cement-based fibre composite. The replacement should not only match the performance of hardened asbestos cement but, importantly, should be also capable of being mass-produced using existing asbestos cement-producing plants, namely of the Hatchek type. AR glass fibres were immediately considered, but could not replace asbestos fibre reinforcement alone because of their very different surface properties (wettability). Different types of 'fibreboard' replacing asbestos cement have been developed and are now in production, some of them using AR glass fibres in combination with other fibres (e.g. cellulose) to permit the large-scale automated production of simple, usually flat elements in Hatchek-type production plants [14].

The beginning of the new millennium witnessed a rapid increase in the production of GRC, fuelled by the construction boom worldwide. The growth was temporarily halted by the post-2008 global economic crisis. However, the industry has now recovered and GRC is already firmly established as a high-tech material of choice in the toolboxes of leading architects worldwide. Designers are becoming increasingly aware of the most recent developments in GRC, and are exploiting them in major projects.

Research into fracture mechanisms and properties of GRC had been for many years restricted to in-house R&D by large companies manufacturing AR glass fibres and licensing their users. Pilkington Glass Co. in the UK did not provide AR glass fibres for any external, independent research for many years. Such attitudes restricted research, adding to the inherent difficulties and challenges associated with the very complex internal structure of GRC. Concrete reinforced with fibres (steel/metal, polymeric fibres, carbon fibres, natural fibres etc.) other than glass fibres had been a popular university postgraduate research topic for decades. A closer look at the plethora of conferences and multitude of publications on 'fibre-concrete' reveals that in many such events, glass fibres were not even mentioned. The situation has improved since the early 1990s, when restrictions regarding access to fibres disappeared. However, current R&D is still confined to a few isolated centres with intermittent projects, recently focusing more on textile (glass fibre) reinforced concrete. Alongside the rapid rise in the use of GRC worldwide over the last decade, there is growing demand for development and advances to produce new versions of GRC of even higher performance. Manufacturers of AR fibres do maintain some R&D in this area. However, coordinated, systematic R&D requires support beyond that which the relatively small/medium-sized companies producing GRC can afford. External, government or trans-government investment and coordination will be required to set up and operate expensive research infrastructure capable of following leads suggested in this publication (e.g. Chapter 6).

Case studies in this book show examples of spectacular effects that have been achieved by using GRC for cladding of buildings and in many other applications. Recent developments include the self-cleaning and photocatalytic (air-cleaning) *e*GRC and panels with smooth surfaces and strong colours. The composite is established in reconstructions of the most complex historic façades and GRC has spread out of construction into the domain of art, where the light-conducting capacity of glass fibres is exploited and items have been produced as interior decorations and furniture.

Constituent Materials | 3

3.1 Binders

A type of hydraulic cement can be used as a binder in the production of matrices for GRC composites depending on how compatible it is with the type of glass fibre to be used as reinforcement. Ordinary cements such as the common varieties of Portland cement (e.g. ordinary Portland cement – OPC), which are the most widely used in construction, produce alkalis during their hydration. Hardened matrices made from such cements also remain strongly alkaline during their service life. Such matrices can therefore only be used in combination with alkali-resistant (AR) glass fibres to guarantee durability. In addition to OPC, two variations of Portland cement that are often used are the rapid hardening Portland cement (RHPC), which differs from the OPC mainly by the fineness of its particles, and white Portland cement, made from raw materials without iron compounds. The latter is preferred where brighter colouring of the GRC surface is required.

Carbon dioxide in air causes carbonation of the surface layer of the Portland cement matrix and a reduction of its alkalinity. In normal service environments such carbonation affects only thin surface layers and has little effect on performance of cement-rich GRC matrices. It is possible to accelerate the natural carbonation of a hardened matrix and quickly render it non-alkaline. This approach has been tried successfully in laboratory conditions [15], but it is a demanding process and has not been adopted into current GRC production practice.

Another approach to reduce the alkalinity of the GRC matrices is the use of pozzolanic additions such as microsilica (silica fume), pulverised fly ash (pfa), ground granulated blast furnace slag (GGBS) and metakaolin. Such additions modify the microstructure and properties of the hardened matrix, and they generally significantly lower its alkalinity. They also affect bond between the matrix and glass fibres, its development with age and alter the fracture mechanisms of the composite.

There is evidence, namely from early developments of GRC, that an adequately performing GRC can be made using non-alkaline matrices [16]. Very low-alkali or non-alkaline matrices are produced using cements of the aluminium silicate and sulphoaluminate type, such as the high alumina cement (HAC), calcium aluminate cements (CAC) and supersulphated cements (SSC). An example of the SSC type of a low-alkali binder based on the sulphoaluminate is the 'Chichibu' cement developed in Japan in the 1980s. Current practice worldwide indicates that the higher costs associated with such very low-alkali matrix materials tend to outweigh savings in the cost of the fibre reinforcement. Significant additional factors may also be introduced, such as the potential 'conversion' and the resulting substantial reduction of strength of the matrix when aged in a warm, humid environment, as in the case of the HAC cement. Mix design and production processes also have to be modified as current admixtures are sometimes non-compatible and curing requirements are different. Commercial use of such GRC compositions is currently small, although significant applications have been reported from China [17].

Properties of any hardened matrix, but mainly that based on OPC, can be modified by a significant addition (10–15% b.w. of cement) of an acrylic polymer [12]. This decreases the modulus of elasticity of the modified matrix overall, making the composite more ductile. It also provides some protection of the fibres from alkali-based chemical attack, but the degree of such protection tends to be variable. Both permeability of the GRC and moisture transfer within it are much reduced, and dense surface finishes are obtained. GRC with such a high content of polymer is rarely used in current practice, compared with the large-scale standard use of a small dosage (typically 5%), mainly for curing purposes.

3.2 Aggregates

The earliest types of matrices for GRC composites were based on hardened cement paste, with little or no aggregate. The early composite was therefore called glassfibre reinforced *cement*. However, it soon became obvious that the cost of the matrix could be reduced and some properties of the hardened matrix could be improved (e.g. reduced drying shrinkage) if greater quantities of fine aggregate were added.

Results from practical applications showed that a significant increase in the aggregate content was very beneficial and the aggregate content of a typical mix gradually increased to the currently prevailing cement/aggregate ratio of 1:1 or more.

The shape of the particles should preferably be round with a smooth surface and no honeycombing. The maximum size of the particles is generally limited to 1.2 mm (spray-up process) and 2.4 mm (premix process). The content of very fine particles (passing 150 μm sieve) is limited to 10% of the total weight of the sample. This requirement tends to exclude ordinary 'building sands', which are likely to have an excess of fines and too many 'oversized' particles. The fine aggregate must be clean and uncontaminated by chlorides, sulphates, alkalis (total <1%) or organic matter (loss on ignition <0.5%) and may have to be washed, if necessary. Silica sands with SiO_2 content >96% are widely used, as their grading is usually well-controlled, the particles are of rounded shape and the aggregate is chemically stable.

Moisture content of the aggregate must be known and controlled. Mix designs are usually based on a moisture content of <2%. Mix proportions have to be adjusted if a higher moisture content is used.

Sands other than silica sands may be used but the producer should provide evidence of their suitability by carrying out production trials and testing of full-scale elements, including an assessment of long-term properties.

Special aggregate may be used for the facing layer if a specific architectural appearance is required, usually achieved by exposing the aggregate in the surface layer at the end of the production process. Aggregate for the exposed aggregate surface is selected for colour or texture, usually from crushed rock. Grading is adjusted to suit the production process selected.

3.3 Fibres

The inherent tensile strength of glass in the form of very thin individual fibres/ filaments (diameter <20 μm) is very high compared with that of the same glass in bulk. It relies greatly on the glass fibres retaining their pristine, undamaged surfaces with very few, if any, defects. In order to exploit such outstanding properties of glass in its fibre form, the production of the fibre itself and that of the composite is designed to maintain the fibres undamaged as much as possible throughout the production process.

Glass fibres are made in a process in which multiple filaments are drawn simultaneously from a container of molten glass. The filaments are immediately coated (a 'size' is applied) and kept together in a bundle. A size/coating is applied as a very thin layer, usually tens of nanometres (10^{-9} m) thick. It protects the pristine surfaces of the freshly drawn filaments from abrasion and keeps them loosely bound in a bundle/strand. The coating influences both the chemical interaction between fibres and the matrix and the mechanical interaction (bond) between the fibres and the matrix, and also between individual fibres within a strand. The coating may also differ according to the production method used. Fibres for the manufacture of GRC by spraying or by premix have a coating designed to enhance integrity of the fibre bundles when sprayed or mixed in.

Less common are fibres with a water-soluble coating, which assists in the break-up of the strands into individual filaments. This is used in applications where a low fibre content is used and the maximum degree of dispersion into individual fibres/filaments is called for. Typical applications of such GRC aim to reduce microcracking or 'crazing' of decorative and facing mixes.

Concrete mixes containing a variety of metallic fibres (e.g. SFRC) may benefit from an addition of glass fibres to enhance resistance to micro-cracking. The same chemical formulation for alkali resistance is used as for ordinary GRC, but the strands have higher internal bonding between the filaments. They are often referred to as 'integral fibres'. The chopped strands have to survive long mixing times and remain as 'solid/integral' bundles even in the presence of high contents of coarse aggregate and usually quite rigid metallic fibres.

Unlike the great majority of other composites with fibre reinforcement, glass fibres in typical GRC production remain largely bundled together. Some of the original strands do break up during the production process into smaller bundles, but few fibres disperse throughout the matrix as single, individual filaments.

It is important to note that the content of glass fibres in GRC is tradition-ally given in % by total weight (mass) of the composite. All the other compo-nents in the mix design for GRC are also usually given as % by weight (mass) of the cement content, unless otherwise stated.

The amount of fibre which can be used depends on the production meth-ods. Spraying or spray-premix processes inherently distribute the fibres mostly in two-dimensional (2-D) layers, which permit a higher packing density of the fibre strands and an optimum glass fibre content of approximately 5% b.w. The premix process usually produces a more three-dimensional (3-D) disper-sion of the strands, with an optimum fibre content of about 3%. Higher-than-optimum contents of fibres for each of the production processes mentioned above reduce workability of the fresh mix, which becomes difficult to compact adequately and reduces its performance when hardened.

Glass fibres are supplied by the manufacturers in different forms:

- *Continuous roving* – the most common form in which bundles of strands with a tex of approximately 2500 are wound on a spool and produce rovings in a cylindrical package containing approximately 20 kg of fibres (Figure 3.1). A typical roving contains 32 strands, which are usually made up of approximately 200 individual fibres. Strands with fewer (e.g. 100, 50) fibres can be obtained.

Figure 3.1 Roving of AR glass fibres.

Figure 3.2 Chopped AR fibre strands.

- *Chopped strand* – strands (Figure 3.2) already pre-cut to a specified length (from 4 to 38 mm) and supplied in bags.
- *Chopped strand mat* – a flat sheet made from lightly bonded short lengths of a fibre strand.
- *Scrims, yarns, nets and fabrics* – used mainly in the production of TRC. However, they can be used as an additional hand-laid reinforcement in the manufacture of large GRC panels, especially when the premix process is used.

3.3.1 Composition and basic properties

AR glass fibres currently available are all based on a silicate glass with a significant content of zirconium dioxide (ZrO_2). The minimum content of ZrO_2 required to provide the fibres with an alkali resistance sufficient to produce a durable GRC has been established at 16%. Current AR glass fibres contain between 16 and 19% of ZrO_2. A typical composition is shown in Table 3.1.

Higher contents of ZrO_2 make the production of fibres increasingly difficult. Attempts have been made to modify ordinary glass through the incorporation of oxides other than zirconia, but no commercial alternative has emerged as a replacement for the existing AR glass [8].

Table 3.1 Typical composition of an AR glass

	SiO_2	ZrO_2	Al_2O_3	CaO	Na_2O	K_2O
%	60–62	16–20	0.3–0.8	0.5–5.6	14–17	0–2

3.3.2 Size, shape and length

AR strands are made from individual filaments with round cross sections and diameters usually between 12 µm and 18 µm. Winding up of the bundle during its initial production causes the cross sections of the strands to flatten.

A typical cross section of a strand embedded in a cementitious matrix is shown in Figure 3.3. Tex of a typical strand, made of 200 filaments, is approximately 75. Reinforcement elements (strands) in the form of a chopped strand are almost straight, with a slight bend reflecting their previous position when being wound onto a spool. Some of the chopped strands, usually the longer ones, may bend more during manufacture, both when being sprayed or cast.

In case of the bond being the same, the longer the strand, the greater its contribution to properties of the hardened composite, and the greater its effectiveness as reinforcement. However, this relationship is a very nonlinear one. Longer strands are more difficult to mix in and to disperse uniformly within the fresh matrix by simultaneous spraying. Compaction of the fresh composite also becomes more difficult with the increased length and content of the reinforcement.

The effect of the length of the fibres on the properties of the GRC reduces as the GRC matures and the bond between the fibres and the matrix increases.

The length of the strands is therefore usually not more than 25 mm in the pre-mix process and 37 mm in the spray-up process. The use of concentric spray-heads permits the length to be increased to 42 mm. Longer strands may be acceptable for production based on an extrusion/pultrusion process. Very long, effectively continuous fibres are used during production by filament-winding. Tensile strength of the glass fibres is maintained up to approximately 200°C. Higher temperatures between 200 and 700°C cause a gradual reduction of the strength. Table 3.2 shows typical physical and mechanical properties of AR glass fibres.

Figure 3.3 A typical cross section of a strand of fibres – shown in a backlit section through GRC.

Table 3.2 Typical physical and mechanical properties of AR glass fibres

| Density (kg/m³) | Modulus of elasticity (GPa) | Tensile strength (MPa) | | Ultimate tensile strain % |
		Single fibre	Typical strand	
2600–2700	70–74	3000–3500	1300–1900	2.0–2.5

3.3.3 Alkali resistance

AR glass fibre was developed for use in cement-based composites [1–4] where the fibres are typically expected to perform for a long period of time in a highly alkaline environment (pH >12.5). Composition of the AR fibres makes them highly resistant to the break-up of the -Si-O-Si- bonds which form the basic structure of a silicate glass. Even the AR fibres are not completely inert when embedded in such an environment for a very long time, and their performance therefore has to be assessed and quantified.

Alkali resistance of glass fibres developed for GRC is best determined directly by measuring the reduction in the tensile strength of single filaments or strands when exposed to an alkaline environment simulating that of a typical matrix, based on OPC. However, fibres would have to be exposed to the exact 'replica' alkaline solution for very long periods, which makes such a test impractical. Moreover, results of direct tests for the 'corrosion' of individual fibres alone cannot be directly correlated with their performance when embedded in the composite in bundles containing varying numbers of fibres.

Many attempts have therefore been made to replace the very long natural exposure time by a shorter exposure of the test fibres to the same solution but at an elevated temperature, usually between 50 and 80°C [e.g. 18]. This approach does generate a stronger attack, aiming to represent a milder attack over a longer period. However, practical experience now available from performance of GRC in service over decades suggests that the alkali-corrosion process at the normal temperature range, even at a very high age, may not reach the results from a much more severe attack during a short test at an elevated temperature. Conclusions from such 'accelerated' tests therefore tend to underestimate the long-term performance of fibres in the GRC. Presence of a polymer in the structure of GRC matrix will also affect performance in an accelerated test using a hot alkaline solution, making it difficult to predict reliably the properties at high age in normal exposure conditions.

A relatively simple test, the 'strand-in-cement' (SIC) test was developed [19] and standardised (EN 15422) [20] in order to enable a quantitative

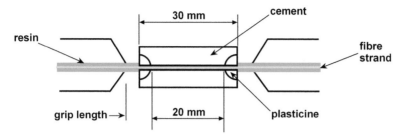

Figure 3.4 Dimensions of a standard specimen for the SIC test [19].

assessment of the alkali resistance of different types of fibres. A standard 'SIC' test specimen consists of a glass strand embedded along the centre-line in a cylinder of the matrix material with a 20 mm contact length in the centre of the specimen (Figure 3.4). Ends of the strand emerging out of the matrix are embedded in plasticine to minimise damage and chances of a tensile fracture outside of the matrix. Both protruding lengths of the strand are embedded in an epoxy resin which permits the strand to be safely gripped. Test specimens are aged in hot water (80°C) for a specified period (96 hours). At the end of the exposure, the resin ends with the embedded strand are gripped in a testing machine and pulled out simultaneously so that the specimen is subjected to a uniaxial tensile load.

The tensile strength of the 'SIC' strand should be greater than 330 MPa in order to be considered adequately alkali-resistant [20–22]. For a test to be valid, the strand must fall within the length embedded in the matrix. Production of test specimens and the test procedure are not difficult, and the test results are widely accepted as benchmark data for assessment of the alkali resistance.

The SIC test is adequate for comparisons between strands of different types of fibres, and on the efficacy of sizes/coating when the chemistry of the fibres is identical.

A nominal tensile strength (based on the ultimate load and total cross section of the fibres in the strand) is recorded. However, the fracture mechanism itself is very complex, and absolute performance data are very difficult to obtain.

The SIC test does not attempt to simulate real GRC; loading conditions differ in that the tensile load is applied directly to the strand, whereas in a real composite it is transmitted from the matrix onto the fibre strands.

Properties of the glass fibre strand as a unit also differ because the bond between individual fibres making up the strand depends on the nature and properties of any material deposited in the inter-fibre space. Type and properties of such interstitial deposits are affected by the elevated temperature of the curing environment during the SIC test.

Degradation of AR glass fibres in a cementitious matrix can be further reduced by an appropriate coating of the fibres. This can provide an effective barrier between the surface of the glass fibre and the alkaline pore solution in the matrix. In addition, a polymer coating of the fibre has the potential to increase its strength by the closing of initial surface flaws [23, 24]. However, the continuing improvement of the performance of the fibres has led to difficulties with the SIC test, as the strand failed too often outside of the embedded length [24].

3.4 Admixtures

Admixtures are substances added into the mix design in small quantities, usually less than 5% by weight of cement. A wide range of admixtures, originally developed for and used in concrete technology, are also used to modify properties of the fresh and hardened matrix, which may also enhance the properties of the composite itself [25]. It is very important to check for any possible effects of the chosen admixture on performance of the final product and on any interactions when more than one admixture is used.

3.4.1 Plasticisers and superplasticisers

The addition of fibres inherently reduces the effective workability of a fresh GRC mix. High workability but stable, fresh slurry is therefore required. Admixtures which improve workability without impairment of properties of hardened GRC (e.g. with no extra water) are therefore in demand. The first type of admixtures were plasticisers, mainly based on lignosulphonates from the waste in paper production. These were superseded by much more effective superplasticisers (such as melamine sulphonates and, more recently, polycarboxylate ethers) which act as very powerful dispersants of particles of cement in a fresh GRC matrix, without significant side effects such as retardation or air-entrainment. They are at present the most common admixtures for GRC. Added at the rates of 0.7–1.5% b.w. of cement, they permit a significant reduction of water/cement ratio and a consequent rise in strength while maintaining or improving workability of the fresh mix. The result is a denser, stronger hardened matrix, achieved often at an earlier age. Such admixtures have become a part of a typical mix design for high-performance GRC.

3.4.2 Viscosity-modifying agents

New types of polymer compounds which influence workability have been developed using molecular design at nano scale. These are often referred to as viscosity-modifying agents (VMAs), which have been specifically developed for self-compacting concrete [26]. Through the interaction of the functional groups of the molecules with the water and the surfaces of the fines, VMAs build up a 3-D structure in the liquid phase of the mix to increase the viscosity and/or yield point of the paste. Such admixtures aim to improve workability of the fresh GRC mix to make it flow and suitable for casting without the need for additional compaction. At the same time, the VMAs enable a very high-workability mix to remain adequately stable to prevent excessive segregation or bleeding during the production processes. Self-compacting fresh GRC enables a faster casting in the premix-type process with no or minimal compaction, usually also leading to better surface finishes.

3.4.3 Accelerators and retarders for setting/hardening

Admixtures similar to those used in ordinary concrete technology are used in GRC production for very similar reasons. The high content of Portland cement in a typical GRC mix generates by itself a considerable amount of heat from hydration, and accelerators in the form of admixtures are not normally required.

Accelerators speed up either the setting or the hardening process, or both. As a result, the heat of hydration of cement is released at a faster rate. Heat produced by their action could be used to counteract effects of low ambient temperatures and to maintain normal curing times. Alternatively, the de-moulding time can be shortened and the production rate increased without the need for a special environment (e.g. steam curing). Accelerators used for spraying concrete (gunite) can be used for instantaneous stiffening of a GRC mix when used *in situ*, such as for tunnel linings. An inexpensive and very effective accelerator is calcium chloride. However, it greatly promotes corrosion of steel, and its use is strictly limited to cases where no steel (ferrous) elements such as fixings are in contact with or embedded in the GRC.

Retarders slow down the setting/hardening process. They are less common in GRC production practice. However, they are considered and used when the composite has to be produced in a very hot environment and adequate means of cooling are impracticable or expensive. If required, admixtures similar to those used in ordinary concrete technology can be used in GRC.

3.4.4 Titanium dioxide

Discovery of the very high photocatalytic capability of nano scale particles of the anatase-type of titanium dioxide [27] has been recently exploited

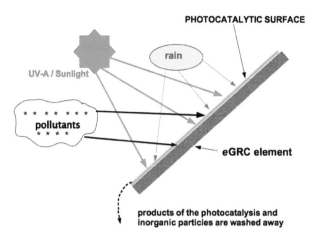

Figure 3.5 Outline of the photocatalytic activity at an 'active' eGRC surface exposed to natural light.

in development of the photocatalytic GRC, the *e*GRC [28, 29]. The first attempts to produce photocatalytic GRC were based on a direct replacement of the binder by a photocatalytic one, in which approximately 5% of cement was replaced by the most photocatalytically active version of TiO_2 (anatase) [30]. However, the TiO_2 becomes active only when the surface of a GRC product is exposed to UV light from natural daylight or from an artificial source. Figure 3.5 shows the de-polluting and self-cleaning mechanism on the exposed active surface of the *e*GRC.

Replacing all of the cement in the GRC matrix by a photocatalytic binder would be uneconomical. The *e*GRC therefore uses the photocatalytic binder only for the very thin, un-reinforced matrix layer ('mist' coat) which is sprayed first into formwork or moulds as a lining. The amount of the photocatalytic matrix required for this purpose is very small, and the cost difference between *e*GRC and an ordinary GRC is reduced to an insignificant level. The first use of photocatalytic GRC was in Italy (Grupo Centro Nord) in 2008. All of the cement used for GRC cladding in this project was replaced by a photocatalytic cement, to prove the basic performance. The first large-scale use of the *e*GRC, where only the surface layer is photocatalytic has been in China in 2011 (Nanjing Beilida Co.). The project is illustrated in Chapter 11.

It should be noted that none of the active TiO_2 is consumed in the photocatalytic process. Unlike many other materials with nanosized particles TiO_2 does not pose any significant health hazard [31] greater than that associated with OPC.

3.4.5 Other admixtures

Pumping aids which are used in standard concrete technology can be also used in production of GRC, namely in the premix process where the fresh GRC is placed by pumping and casting.

Rarely used in GRC are other admixtures for concrete, such as those aiming to reduce permeability of hardened concrete.

3.5 Polymers, pigments and additives/fillers

3.5.1 Polymers

The addition of a polymer compatible with cement has an effect on both fresh and hardened GRC, its significance depending on the type of polymer, the amount used and curing conditions. Workability of fresh mixes is usually improved. Properties of hardened GRC are influenced via the effect of the polymer on bond between fibres and the matrix, and on the modulus of elasticity of the matrix. The latter lowers noticeably the overall modulus of elasticity and increases the ductility of the composite when added in greater quantities (approximately 8% and more). Smoother and denser surface finishes are possible, compared with ordinary GRC. Increase in the density of GRC also makes it less permeable to water, which improves its durability overall. However, on account of the polymer being a carbon-based compound, the hardened composite will no longer be considered as entirely non-combustible. However, other key fire properties remain unaffected.

Acrylic thermoplastic polymers are commonly used, as they are stable in the alkaline environment of a typical cementitious matrix of GRC. They are typically added during the mixing process, and an allowance for the water content is made in the mix design. Polymers are normally used in the form of an aqueous dispersion containing between 45 and 55% of solid matter and powder forms with a solids content of 90–100%. Typical water-based polymer dispersions are sensitive to frost damage in storage, and in order for the polymer dispersion to retain its film-forming capacity, temperature of the fresh mix during manufacture should not be lower than 7–12°C, depending on the specific polymer used.

The effect of the acrylic polymer on the hardened GRC depends largely on the quantity used. Small quantities of acrylic polymers in the form of powder, dosed at around 1% by weight of cement, or in an equivalent liquid dispersion are still classed as admixtures, and are used principally to eliminate the need for the wet curing of freshly made GRC. Such quantities of polymer do not have a strong direct effect on durability and mechanical properties of hardened GRC. However, indirectly, by enabling adequate curing of the fresh GRC, they

Table 3.3 Specification for acrylic polymer admixtures (for curing) [32]

Compound type	Aqueous thermoplastic polymer dispersion	Redispersible thermoplastic polymer powder
Polymer type	Acrylic based	Acrylic based
Solids	45–55%	90–100%
Appearance	Milky-white, creamy, free from lumps	Free-flowing white powder
Minimum film-formation temperature	7–12°C	0°C
Ultraviolet resistance	Good	Good
Alkali resistance	Good	Good

do improve the quality and long-term performance of the hardened composite overall. They should be used in accordance with the manufacturers' instructions, and should conform to the specification in Table 3.3 [32]. Polymers with properties outside the above specification may be used with the agreement of the purchaser, supported by adequate test results.

3.5.2 Pigments

Coloured GRC is normally 'coloured-through'. This is generally preferred to a GRC with a painted surface. Any type of pigment, both in powder form or as a water-based suspension, can be used provided it is compatible with the cementitious matrix and its alkaline environment. It is difficult to predict accurately the colour of the final hardened GRC. Production of trial mixes and prototype elements is therefore essential in order to agree on the colour and how much variation in its shade will be acceptable. Coloured GRC requires a particularly high level of control of the production process to ensure a uniform distribution of the pigment within the body of the composite. Surface treatment (staining) may have to be used if a very high level of colour consistency is required.

Production and use of coloured GRC have become very common, and are used to a great effect (see Chapter 11). In order to produce light-coloured surfaces, or to enhance the brightness of colour of the pigmented surfaces, white Portland cement is widely used as the binder. Recent development of the photocatalytic *e*GRC suggests that it can also be used to enhance and maintain for longer the brightness of the surface.

Very dark external surfaces tend to be avoided. A dark colour greatly increases the heat gain from sunshine, with surface temperatures potentially higher than 65°C. External GRC elements, such as cladding, then become very hot and develop excessive thermal expansion and movement beyond that which can be accommodated by fixings. A very high temperature gradient between the external and internal face of a thin-walled element combined with internal stress generated by the restrained thermal movement are likely to lead to excessive distortion/bowing and cracking. In addition, very dark surfaces are more difficult to replicate with very low colour variations, and tend to show any efflorescence occurring in the early age of GRC exposed to wet weather.

3.5.3 Microsilica

Microsilica are very small (<1 μm) particles of amorphous silica, originally a by-product from production of ferro-silicon. The silica in this form is highly reactive with calcium hydroxide in the cementitious matrix of the GRC. The very small size of the particles and their reactivity are already exploited in ordinary concrete technology, where microsilica enhances strength and acts as a densifying material, improving some aspects of durability.

Attempts were made at modification of the GRC matrix by replacement of up to 10% b.w. of cement by microsilica. Such a modified matrix led to a higher strength and enhanced durability of the composite. Investigation of the fracture mechanism of GRC made with glass strands impregnated by microsilica indicated [33] that microsilica largely prevented crystallisation of $CaCO_3$ between the filaments of a strand and maintain longer the desirable 'telescopic' mode of fracture. However, adding microsilica complicates GRC manufacture, and a reliable pre-impregnation of the glass fibre strands is technically demanding and costly. This explains why microsilica is not used in current general GRC production practice.

3.5.4 Hydraulic additives

Several types of hydraulic (pozzolanic, reactive) materials can be used for a partial replacement of cement (see Section 3.1) in the GRC matrix. The most common materials in this category are pulverised fly ash (pfa), metakaolin and GGBS. Both the pfa and the GGBS are occasionally used to save on the cost of cement. However, the saving in cement content is usually accompanied by a lower rate of the strength gain of the composite, and curing regimes may have to be altered to promote an adequate hydration of binders with such additives.

Metakaolin is a form of active clay which reacts with lime liberated during hydration of cement in the GRC matrix; it reduces the cement's alkalinity.

The process also reduces the formation of lime crystals in the interstitial spaces between fibres inside a fibre strand, which modifies the bond between the inner fibres. GRC based on a metakaolin-type matrix shows an improved long-term performance, namely in the retention of flexural strength and toughness. This is likely to be a reflection of the mode of fracture being more of the 'telescopic' or gradual/combined fracture type. The surface finish of GRC with metakaolin is improved by a reduction of efflorescence, when exposed to weather. Unlike the pfa and GGBS additions, which also considerably darken the surface of a finished product, the addition of metakaolins only results in a slight creaming of white cements.

3.6 Water

Quality of water must comply with what is required for ordinary concrete [34]. Water should be clean and free from deleterious matter.

Manufacture | 4

4.1   Batching and mixing

Batching methods reflect the nature of each of the constituents and the production method chosen. Weigh-batching applies to basic ingredients of the matrix, with specialist dispensers used for adding precise amounts of small quantities of admixtures and additives. The weighing equipment for solid ingredients must be calibrated and capable of accuracy within ±2% of the stated target batch weight. Liquids are measured by weight or by volume, and they are often dispensed automatically, with the same accuracy as for the solids.

The input of glass fibres is usually controlled by the size/number of rovings and the speed at which they are fed into the chopping mechanism when the spray-up production method is used. It is essential to control the fibre content of GRC very closely, and the mechanism feeding the fibres must be accurately set. Performance is monitored continuously by a fibre output rate monitor, or the machine needs to be regularly tested for the accuracy of its delivery rate of the fibres and the slurry. The 'bag and bucket' test (Sections A.1 and A.2) is used and the settings adjusted.

The premix production method uses weight-batching for all constituents charged into an appropriate mixer. Most of the common mixers used are designed specifically for glassfibre reinforced concrete production. They are of a forced-action type, with a batch size of approximately 0.060 m³ (60 l) or 125 kg, the size being based on the use of bagged cement charged directly into the mixer. Larger batch mixers and continuous action mixers are used in integrated batching and mixing plants capable of a high output of GRC of a variety of mix designs.

Large batch sizes are generally not beneficial for sprayed GRC because of the rate of usage. It is preferable to make small (approximately 125 kg) batches, which are completed in 10–15 minutes. Modern GRC mixers can produce such

Figure 4.1 Typical GRC 125 Combination high shear batch mixers. (Courtesy of L. White.)

a batch in under 5 minutes. This is of particular importance when producing in hot climates, where the workability of the fresh mix reduces rapidly.

Premix mixers do sometimes have larger capacities, in particular when mixing self-compacting premix GRC mixes. However, the workability of these mixes also reduces rapidly with time, and hence batch sizes tend to be limited to less than about 250 kg depending on the time required to use the batch and the climate where the factory is located.

A typical mixer for the spray-up process is capable of generating a high shearing action using mixing tools rotating at a high speed (Figure 4.1). Such mixers are usually operated in two modes/stages: first, a high-speed (high-shear) action produces a high-workability slurry for the spray-up production; second, this is followed by a period of low-speed (low-shear) action for the blending-in of pre-cut fibre strands for the premix GRC production.

Mixing time for one batch is usually between 60 and 120 seconds. Longer mixing times do not improve workability any further and may lead to excessive entrainment of air and increased temperature of the mix.

Other types of mixers can be used. These include a 'continuous' rather than the typical 'batch' mixer, or a mixer with a flexible drum, suitable for blending different types of fibres into fresh concrete. Small-sized standard concrete mixers are also used, provided they are capable of producing fresh GRC of an adequate quality.

It is possible to have separate batching, mixing and placing/spraying equipment; however, the highest productivity and often the highest uniformity/quality of the GRC are achieved when an integrated system is used (Figure 4.2).

Figure 4.2 Integrated automated batching/mixing plant. (Courtesy of L. White.)

4.2 Production of GRC elements

The production process governs the formation of the internal structure of the GRC; it is therefore of fundamental importance regarding the performance of the hardened composite. The properties of the hardened GRC depend very significantly on the degree of compaction of the freshly made composite, which is inherently a porous material. Some of the GRC production processes already provide a degree of compaction, such as the compression applied during an extrusion process or 'pressing' of elements and when a fresh GRC mix has such a high workability as to make it self-compacting when placed by casting.

A common production method is simultaneous spraying, which inherently generates a degree of compaction of the fresh GRC by the impact of the mix hitting the mould. However, an additional compaction has to be applied, usually a manual one using handheld rollers (Figure 4.3a, b).

It is essential to ensure that all of the GRC is compacted, and special care needs to be taken regarding corners in elements of complex shapes. Compaction of the premix GRC is usually achieved by the mix being hand-packed or trowelled into a freshly filled mould, after which the whole content and the mould are vibrated. The requirement for compaction is eliminated when a very highly workable, flowing, self-compacting mix is used.

It is important to understand that the type of the production process also has an effect on the orientation of the fibres. It determines the degree of

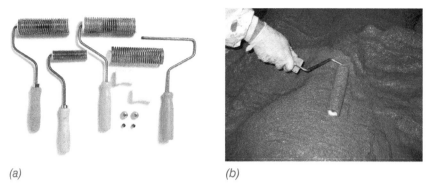

(a)   (b)

Figure 4.3 (a) Spring rollers for manual compaction of fresh GRC; (b) manual compaction of a highly profiled GRC element using a roller. (Courtesy of L. White.)

anisotropy of the hardened GRC produced. Care must therefore be taken to ensure that all of the GRC is compacted.

Spraying tends to arrange most of the fibres in a two-dimensional (2-D) random distribution, parallel to the flat surface of the mould and 'in-plane' of the sheet/panel produced. The length of the fibres usually significantly exceeds the thickness of GRC elements produced by spray-up. A discernible 'wall effect' can be generated, which reduces the overall fibre content in the composite near to mould faces. In addition, the relative proximity of the mould face further reduces the number of fibres remaining at or near to a position perpendicular to the plane of the element, leading to more of a random 2-D rather than a 3-D distribution of fibres. Such a 2-D distribution can have a positive effect in that it tends to improve the flexural strength of a hardened, flat, sheet-like GRC element.

On the other hand, there are then very few, if any, fibres in positions in which they contribute to transverse tensile strength or to inter-laminar shear resistance of the same flat-sheet element. This explains the low level of transverse tensile strength and inter-laminar shear resistance of typical flat-sheet GRC products. Tests for properties such as flexural strength are normally carried out on specimens positioned either 'mould face' up or down in relation to the applied load (see Section 7.4.2 and Appendix C).

Earlier production methods (see Section 4.2.2) used twin-head spray guns, which, if incorrectly used, could produce non-uniform distribution of fibres 'in-plane' of the sheet, making the 2-D distribution non-uniform and decreasing the characteristic flexural strength of the composite. Earlier testing procedures therefore included testing of not only specimens in two positions regarding the load and the mould face, but also specimens cut out in directions differing by 90°. Modern concentric spray heads have practically eliminated this type of anisotropy, and the 90° 'orientation' tests are rarely carried out.

Extrusion processes, which have been tried on a small scale, also inherently affect the orientation of the fibres in a fresh GRC mix. As expected, the fibres align mainly along the direction of the extruding path [35]. The anisotropy of the hardened GRC produced in this manner may be less than that of the sprayed GRC; however, this is due to the fibres used in extrusion usually being significantly shorter than those in the spray-up process. The length of fibres in extruded GRC can be even shorter than the thickness of the GRC element produced, but in this case, the efficiency of the fibres as reinforcement may be reduced.

4.2.1 Premix and casting

In this process, a cementitious matrix is produced first (hence the 'premix'), and between 2 and 4% b.w. of pre-cut alkali-resistant (AR) glass fibres in the total mix are then blended in. The length of the pre-cut fibre is usually 6 or 12 mm, but 18 and 25 mm and other lengths are also available. As in the sprayed GRC process, the matrix is usually produced in a high-shear mixer in order to obtain maximum workability, and the chopped fibres are added in the second stage, often in a low-speed mixing regime. This approach facilitates their dispersion at the highest practical volume with a minimum of damage to the fibres.

The resulting mix can be hand-packed or trowelled into a mould or cast-in. Casting is either accompanied by vibration of the whole element or without vibration, if a self-compacting mix is used.

The fresh premix GRC can also be sprayed in a process known as *sprayed premix* (which differs from the *simultaneous spraying* described below) using a specially developed spray gun (Figure 4.4). Key items of equipment are an appropriate, preferably two-stage, mixer, a pump for delivering the fresh material and the spray gun designed for an already premixed fresh GRC. Spraying generates a degree of compaction in the process itself, and it is normally not necessary to apply an additional external compaction.

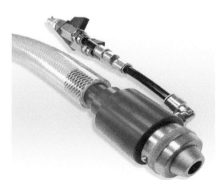

Figure 4.4 Spray gun for the premix spray process. (Courtesy of L. White.)

The method is particularly suited to the manufacture of large numbers of standard products. A current production plant is capable of production of up to about 2 tonnes of such a fresh mix per hour.

4.2.2 Simultaneous spray-up

One or more rovings of glass fibres are fed continuously into a concentric spray gun (Figure 4.5), where the strands of fibres are chopped to a pre-set length, usually between 25 and 37 mm, and projected out of the nozzle by compressed air.

Simultaneously, a freshly made cement/sand slurry is pumped to the nozzle, where it is atomised and sprayed out, mixing with the chopped strands. The resulting fibre and slurry mix is sprayed from the concentric spray-head onto the surface of a mould.

The chopping rate of the fibre and the pumping rate of the cement:sand slurry are typically set to produce GRC with 4.0–5.0% of fibres b.w. of total mix. It is difficult to 'mix-in' fibre contents greater than this in the spraying process and to ensure good compaction when placed.

A 'mist coat' is applied by spraying as the first layer, as thinly as possible, to the surface of a mould. Subsequent layers of fresh matrix mixed with fibres are then applied without any delay to ensure complete integrity. In the case of a 'facing mix' being applied, either by spraying or by pouring, the layer of the facing mix is usually allowed to stiffen before the bulk of the fresh GRC is built up by spraying further layers. The facing layer is usually 3–5 mm thick, depending on the type of surface treatment. Any delay between spraying of successive layers should not exceed the initial setting time of the mix.

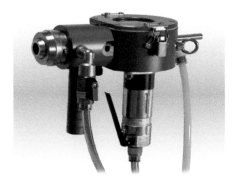

(a)   *(b)*

Figure 4.5 (a) Concentric spray gun; (b) in action. (Courtesy of the iGRCA.)

Figure 4.6 A thickness gauge being used to measure the thickness of a GRC sheet during production. (Courtesy of L. White.)

Sprayed GRC is manufactured in layers. Each complete pass of the spray gun deposits a layer approximately 4–6 mm thick. A typical 12 mm thick panel therefore requires two to three complete passes.

Care has to be taken to ensure that an adequate thickness of the GRC is produced in corners and edges of moulds with complex shapes.

After each layer is sprayed, the composite is compacted with a hand roller to ensure that the panel surface will conform to the mould face, help remove entrapped air and aid the bonding of the glass fibres to the cement paste. The thickness of the product has to be checked after the final pass using a depth gauge (Figure 4.6) or a template. The 'back face' is then finished as required either by the compaction roller or by a float.

Thickness stated in a specification for GRC is the minimum thickness, which must be achieved in all parts of a product. It is expected that in parts such as corners and where the panel is deeply profiled and around anchorage points, the thickness will be greater than the specified minimum.

Thickness of flat areas should not exceed the specification by more than 4 mm. Care has to be taken to ensure that the weight of the completed product does not exceed the maximum weight assumed in the structural design. An over-spray may require a careful removal of the excess material, which has to be discarded, adversely affecting the economy of the production process.

4.2.3 Other production methods

Filament winding

Filament winding consists of wrapping an AR glass fibre reinforcement, impregnated with fresh matrix material, around a mandrel, which defines the final shape of the finished product. The impregnation of the continuous glass fibre strands is done by pulling them through a bath full of a modified Portland cement matrix until 'fully wetted out'. The nature of the process limits the range of shapes of elements that can be made, but it is very suitable for hollow tubular products.

The winding process is computer aided with electronic/electrical components. The computer is programmed to control a closed-loop system, controlling servomotors, which coordinate the winding, pulling and guidance of the composite products. This ensures that the mandrel is wrapped with the required number of matrix-impregnated strands in an optimum orientation to produce a reinforced structure with specific designed properties.

The glass fibre reinforcement used is generally in the form of a single end, tensioned roving, with a number of rovings being placed on a creel and fed through a guiding and tensioning system into the impregnating bath.

The end product is mostly hollow, with the process delivering a high strength-to-weight product. The manufacturing process is similar to that used extensively in the fibreglass-reinforced plastics industry. A vertical filament-winding process has been used in practice for GRC poles and pipes. The resulting composite consisted of a highly modified Portland cement matrix reinforced with up to 18% volume fraction of continuous AR glass fibres. The vertical mandrel rotated on its axis while multiple strands/bands of fibre were placed on the mandrel to achieve the required degree of reinforcement and the desired performance characteristics.

After completion of the winding, the compaction process proceeds along the entire length of the mandrel, improving the 'wet-out' and the alignment of filaments to ensure a proper preparation for curing. The product is then wrapped in a tensioned cover before it is cured for up to 24 hours. The final step in the winding process is to remove the mandrel from the finished product and cycle the mandrel back into a new production sequence.

The flexural strength (MOR) of GRC made by this process can be greatly increased; from the typical value of 25 MPa for an ordinary sprayed GRC it can reach 100–200 MPa, or more, in extreme cases. The very high MOR opens the way for additional structural applications of GRC. Transmission poles made from filament-wound GRC have also shown excellent durability in accelerated ageing tests [36] including exposure to UV, moisture and freeze-thaw cycles.

This method is already used in the USA, Australia and New Zealand in the commercial production of distribution poles for power distribution, solar poles, lighting poles and marine pilings.

Extrusion

Panels and similar elements in which one or both faces are to be smooth, and where the elements are required in large numbers with very high repetition, can be produced by *extrusion*. Small-scale trials have been carried out [35], but the process has not been adopted in general GRC production practice. The main reasons are the cost of the equipment and uncertainty about its adequate utilisation.

Small-scale, mainly laboratory-based, trials have shown that it is also possible to produce GRC elements by an *injection moulding* process [37].

Manual/hand lay-up

Flat GRC boards/sheets and simple 3-D shapes can be produced by this process. Typically, a mould is coated with plain GRC matrix, and it is then lined with a chopped strand mat, an AR fabric or a mesh. Another layer of fresh matrix is applied, penetrating the fabric, and the next layer of the mat, or of a glass fibre fabric, is placed. The process is repeated until the required thickness is achieved. The process is very close to that of making elements out of textile-fibre-reinforced concrete (TRC). It is also possible to combine premix process with a hand lay-up, using a glass fibre textile mat as additional reinforcement in vulnerable parts of the product.

A similar technique, in principle, is used to produce *light-transmitting concrete* (Figure 4.7). One approach is to produce a bulk element with layers of parallel glass strands first and then cut out 'slices' very precisely from the bulk element, with fibres perpendicular to the plane of the cut [38].

The production of GRC elements using *3-D printing* has been developed very recently in China, pioneered by the WinSun Co. [39]. It is currently at a

Figure 4.7 Section through light-transmitting glass fibre concrete [38].

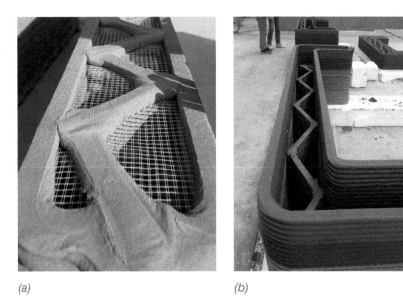

(a)   *(b)*

Figure 4.8 Details of lattice truss elements produced by 3-D printing of GRC. (a) Individual 'truss-like' elements are printed first. (b) The elements are stacked and the inner void filled to form a wall unit. (Courtesy of Yanfei Che.)

stage in which thin 3-D elements are 'printed' and then joined up by stacking, or tilted up and connected, to make a structure (Figure 4.8). Complete buildings have been assembled in this manner. Another successful trial has been in 3-D printing out and stacking horizontal 'slices' of the structure made from GRC, which created voids for traditional reinforced concrete, effectively serving as a permanent formwork.

4.3 Curing

The length of the curing period and the curing conditions have a significant effect on the properties of hardened GRC. The basic requirements are similar to those applicable to concrete technology in general: mainly, the prevention of the loss of water content, the drying-out of the freshly made product and the maintenance of an adequate temperature. However, considering that GRC has a higher cement content than that of an ordinary concrete and tends to be used in the production of elements with thin cross sections, the curing regime for GRC has to be controlled much more accurately than in the case of ordinary concrete.

The curing regime adopted must ensure that the composite attains adequate strength by the time it is to be removed from moulds/formwork and handled, and that early drying shrinkage is minimised.

Curing conditions are determined by many factors. The principal ones include:

- the type of cement;
- the mix design overall; and
- additional requirements regarding productivity/economy.

Early production processes required freshly placed GRC to have any excess water removed first by de-watering, usually by vacuuming out water via a porous suction membrane placed on its surface. The remaining water in the mix was then 'sealed' within the element by wrapping it in a tightly fitting polythene sheet and then cured in a warm 'fog room'. Alternatively, the freshly made element was immersed in water. Such a curing process complicated the production of GRC, and it is no longer used.

Curing remains one of the key processes in modern production systems of GRC, and the provision of a moist environment is still a fundamental requirement. However, current production processes have dispensed with most of the early measures described above, mainly by a change in the mix design. This has led to both productivity and economy being much improved. It has now become routine to add small amounts of acrylic polymers (see Section 3.5.1) into the fresh mix to achieve adequate curing. The added polymers keep the internal moisture in and prevent its undesirable loss by evaporation from the fresh composite during its curing period.

The curing environment still has to be controlled, ideally at approximately 20°C and 95% RH. Any sudden and rapid drying-out or large temperature changes must be avoided before the GRC reaches adequate strength for the product to be safely removed from a mould and for the days immediately afterwards. This is particularly important for the first 2 days after manufacture of GRC containing polymer, when the ambient curing temperature should not be below 5°C or above 35°C.

4.4   Moulds and formwork

GRC has an exceptional ability to enable the manufacture of products of very complex shapes. Exploitation of this capability therefore requires very high-quality moulds/formwork. The performance requirements for moulds and formwork for GRC tend to be greater than those applied in ordinary concrete technology, although the overall mass of material placed and the formwork pressures are lower in the case of GRC. The moulds for GRC can therefore be much larger and very much more complex and can be reused more often than those for ordinary concrete.

Greater accuracy and tighter dimensional tolerances are also required. The moulds have to be designed with an appropriate allowance, corresponding to the expected volume changes of the GRC element, such as those due to shrinkage or changes in temperature. Care has to be taken to ensure that acceptable dimensional tolerances are not exceeded.

It is also essential to consider the movements/dimensional changes of the fresh composite and avoid any such movements being restrained by the shape of the mould during the curing process. Any such 'self-restraint' of the inherent, natural movement (e.g. shrinkage) will generate internal tensile stresses. These may lead to cracking and an impaired overall performance of the GRC element in service. If possible, corners should be slightly rounded or chamfered and very sharp angles avoided.

Mould release agents used in concrete technology are also used for GRC. The agents should not interact with the surface of the mould or with the GRC.

Rigid moulds made from one type of material (timber/plywood, steel) were the prevailing type in the early years of GRC. Fibreglass (FRP) moulds were then shown to be easier to make when products with complex shapes were required. Two-component moulds, in which a rigid base is lined with another material, such as timber lined with a polymer resin or steel lined with polyurethane rubber, are now common in current GRC production. Such moulds and mould liners exploit the development of silicone rubber and polymeric elastomers [40]. Polyurethane rubber moulds are now particularly popular due to the ease of stripping of GRC elements with very complicated surface patterns, minimising breaks and damage (Figure 4.9). Such moulds are also capable of being reused many times, and they produce elements even with very deep textures and undercuts.

Figure 4.9 Polyurethane rubber mould being stripped off. (Courtesy of the iGRCA.)

Figure 4.10 Mould for production of a prototype GRC panel forming a section of a large dome. (Courtesy of the iGRCA.)

The design of an appropriate and effective mould is very important. It is usually necessary for a full-scale model of the element to be constructed, such as that shown in Figure 4.10, and the mould is 'cast' off it, observing appropriate dimensional tolerances. The model is usually treated with a release agent to facilitate the removal of the freshly set mould.

It usually takes 3–7 days for the mould material to cure and retain its dimensions accurately. The mould must be shaped in such a manner that it can be lifted or stripped easily and without damaging the finished GRC product.

The recent architectural trend to use curved façade elements for flowing, free-form shapes of buildings, for which GRC is very suitable, generated new challenges for the mould-making process. The latest developments have integrated the digital design of façade elements with new mould-forming technology [41]. The new process (Figure 4.11) substantially reduces the amount of manual work required to produce moulds with curved surfaces (Figure 4.12).

4.5   Surface finishes and treatments

The surface of the mould used is reflected in the initial surface finish of the GRC product. It is therefore essential that there are no defects or visible joints in the surfaces of the moulds. The original surface of the GRC can be subsequently treated to produce an even greater range of finishes. Surface finishes vary from very smooth and glossy to flat matt and a wide array of textured and patterned surfaces with or without pigments.

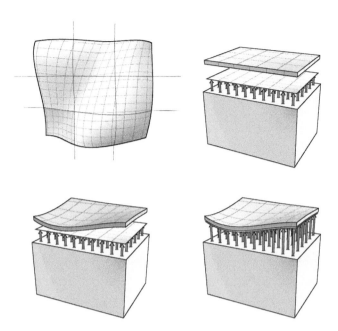

Figure 4.11 Principle of the 'adaptive mould' system for production of moulds for double-curved GRC elements. (After Raun and Kirkegaard [41].)

Figure 4.12 Production of a GRC element using the 'adaptive' mould process. (After Raun and Kirkegaard [41].)

GRC can be made with either

- *An ex-mould finish, either smooth or textured, according to the mould surface*

    Smooth and polished moulds or mould liners produce equally smooth, even glossy, GRC finishes. Such finishes tend to be sensitive to hairline crazing of the surfaces. Any blemishes, blowholes and imperfections become more visible compared with other finishes. If a mould is used with a textured surface, a mirror image of this surface will be faithfully reproduced onto the finished GRC product, also making minor imperfections less visible.

or

- *An exposed, fine aggregate finish (with or without a texture)*
  In addition to textures and patterns that are generated by spraying or casting into appropriately shaped moulds, sometimes with liners, surfaces of GRC can be also treated chemically or mechanically either when the product is freshly de-moulded or when it is already cured.

Water-washing away a freshly set, but not hardened, cement paste is sometimes assisted by applying a coating of a retarder onto the mould surface prior to the facing mix being applied. In a typical treatment, the thin top layer is removed by a water jet to reveal a specially selected decorative fine aggregate (the 'exposed aggregate' process) on the external surface of a panel. Washing out by water is sometimes augmented by a treatment with a weak acid (usually hydrochloric) to dissolve the thin top layer of a cementitious paste, which would have obscured the decorative surface of the exposed aggregate when the emphasis is on colour and texture.

Strong acids can also be applied to the de-moulded surface to expose the decorative sand or fine aggregate. This is usually done immediately after de-moulding, as the surface of GRC rapidly becomes impervious and difficult to penetrate. Depending on the intended visual effect, the washing out varies from light and shallow to deep, with greater exposure. The former reveals the finer aggregate, and the appearance and colour of the sand will prevail on the surface. The latter exposes more of the coarse aggregate, which will then determine the appearance. Surfaces are wetted before the acid treatment to prevent the acid penetrating more deeply than required. It should be noted that the very 'bright' appearance of etched aggregate will tend to fade with exposure in service, particularly when a calcareous stone is used. Acid washing can be repeated for a deeper exposure, but the duration of the wash has to be consistent and the surfaces washed by freshwater immediately after each treatment.

Mechanical treatments aim to produce a surface with a texture simulating natural materials, such as either a rough or a polished stone. This can be achieved by wire-brushing and grit-blasting or grinding/polishing at a suitably early age of the composite. Grinding and polishing treatments can only start after the GRC has adequately hardened (>35 MPa compressive strength). Insufficient strength will cause particles of aggregate to be dislodged and damage the surface. Grit-blasting should use materials that do not fragment and produce excessive amounts of dust and do not discolour the GRC surface. Silica sands used for production of GRC are suitable for this purpose. Care must be taken to ensure that the depth of the ground layer is consistent and that several stages are used rather than a single very deep 'cut'.

Surfaces of GRC can be painted. The paint is usually applied onto a pre-pigmented GRC of a similar base colour. There are a number of suitable proprietary paints and penetrating stains now on the market that can be used to finish the appearance of a GRC product. As with all surface finishes, the manufacturer's instructions must be strictly adhered to, and the product must be compatible with the GRC substrate over a longer period.

Special surfaces can also be created by using GRC as a substrate carrying an attached 'veneer' of a decorative material (ceramic tiles, sheets of terracotta or slices of clay bricks, polished stone, etc.). This is a highly skilled operation, which may require special fixings and adhesives, and great care has to be taken in observing the veneer suppliers' recommendations.

Hydrophobic coatings and surface sealants used for ordinary concrete can be applied and must be adequately permeable to water vapour. Such coatings have the potential to reduce lime-based efflorescence on the external surfaces and the effects of a polluted environment. Coatings of this type are not frequently used in current practice, and verification of the compliance of a completed treatment on a GRC element with the specification for the treatment can be difficult.

The production of surface finishes of consistent appearance requires high levels of skill and supervision. Special attention has to be given to potential differences in volume changes, namely drying shrinkage, between the bulk of the GRC and the modified surface layers. This is particularly important in the case of GRC with a veneer of a different material.

In all cases, trial elements with typical surfaces of an agreed standard should be kept throughout a project to provide continuous benchmarks for acceptable variations in the quality and appearance of the selected surface finish.

4.6 Handling, transport, storage and repairs

Thin-walled GRC, particularly very large panels and other freshly cured products, remains vulnerable to permanent damage if not handled, transported and stored in an appropriate manner.

The handling of a GRC panel, including the effect of an asymmetrical lift, should be considered as part of the structural design. It is essential to use only the lifting points identified in the design, and special lifting frames are used for handling large panels (Figure 4.13).

If the units are stored stacked, all the points of support must be vertically aligned and in the correct locations. *Embedded fixings can be used for lifting GRC elements only if they were also designed for this purpose.*

Being lightweight, GRC products are often transported over considerable distances between specialist production plants and construction sites, the low weight being an advantage over precast concrete and other construction systems. However, thin-walled GRC elements are prone to impact damage, and care must be taken to minimise any vibration and to restrain their movement during transport. Anchorage points should be provided for straps to hold the units safely attached to racks or transport frames, and the units must be separated by a soft packing. Adequate additional protection must be provided for any particularly vulnerable parts, such as corners. The transport of very large GRC panels requires frames/racks specifically designed for this purpose.

Incorrect storage may lead to permanent distortions and unacceptable deflections/deformations and thus lead to a rejection of otherwise perfect products.

Any damage to load-bearing GRC elements must be assessed from the view of structural performance. If not structurally significant, minor damage

Figure 4.13 Handling of a large GRC panel. (Courtesy of Nanjing Beilida Co.)

can be repaired using fresh GRC mix similar to that used for the production of the element (see Section 9.7).

External protection is often used to ensure that the exposed surfaces are not accidentally stained. Care should be taken to avoid freshly made GRC being stained or collecting dirt. Stains from manual handling can be removed by using household detergents and washing with water.

4.7 Cutting and shaping

Hardened GRC can be cut using standard equipment for concrete, such as diamond-tipped saws and drills. Recent development includes precision cutting with the lowest amount of waste by the use of laser beams (Figures 4.14 and 4.15).

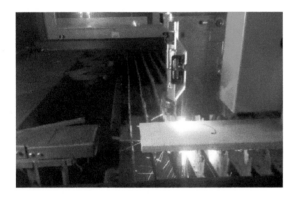

Figure 4.14 A laser beam cutting through a sheet of GRC. (After N. Charlesworth.)

Figure 4.15 A GRC sheet with a very precisely cut hole using a laser beam. (After N. Charlesworth.)

Composite Action | 5

5.1 Internal microstructure

The microstructure of the fresh GRC composite is determined largely by the composition of the mix and the effects of subsequent mixing and production processes. However, in common with all composites based on hydraulic cements, the initial microstructure continues to change and develop with age in a non-linear manner and with a high dependence on the service environment.

The cross section of typical GRC in Figure 5.1 shows the primary reinforcement: bundles of glass fibres embedded in a matrix. The composite is typically not only made up of the solid components of the mix, but also contains voids filled with water and/or air.

5.2 Fracture mechanisms

Internal stresses, including tension, develop when an external load is applied to a GRC element. Tensile stress becomes critical stress when it exceeds the tensile strength of the matrix and causes it to crack. The mode of fracture (failure mechanism), which follows the first cracking of the matrix, varies significantly depending on the properties of the matrix and fibres and on the bond between them [42]. As the properties of the constituents and their interactions change with age, so does the mode of fracture.

The mechanical interaction between a fibre strand and the matrix in which it is embedded is measured by the strength of their bond. In addition to bond between fibre strands and the matrix, the mode of fracture of a GRC element also reflects bond between individual filaments within a strand. The passing of time – ageing – and conditions of exposure affect to a different extent both the

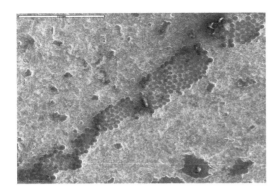

Figure 5.1 Cross section of GRC showing strands of glass fibres. Note that almost all of the fibres remain in bundles of different sizes.

strand–matrix bond and bond between fibres within a strand. Bond between fibres within a strand determines the degree to which the strand acts as a single 'solid' element or as a 'composite' fibre bundle.

Research [42–44] has shown that three basic types of failure mechanism follow once the first crack occurs in the matrix; this occurs at around the level of stress corresponding to the limit of proportionality (LOP). At this point, the tensile strength (or strain capacity) of the matrix is exceeded by the internal stress due to the load and the matrix cracks. At this point, the matrix ceases to carry any stress, and the stress, until then taken by the matrix, is instantly added to that already taken by the strands of fibres crossing the crack. How much of the stress an individual strand crossing the new crack can take depends on many factors, such as its embedment length, its angle in relation to the direction of the stress and, of course, its peripheral and internal bond.

A further increase in load (stress) leads to an eventual failure of the strand, and provided the plane of the crack is largely perpendicular to the longitudinal axes of the strands, the fracture mechanism falls into three general categories (see Figure 5.2).

1. *Complete pullout of the strands*
   This occurs when bond between the fibres at the perimeter of the strand and the surrounding matrix is low, but bond between fibres within a strand is high. It enables the whole strand to act as a single, integral reinforcing element.

The tensile strength of the inner fibres, which make up most of the strand, and their reinforcing capability are therefore poorly utilised. The strands tend to pull out as single elements. This is seen in Figure 5.3, which shows the fracture surface of a flexural test specimen from a premix GRC. The great majority of the strands pulled out instead of breaking in tension. The work of fracture was low. This mode of fracture is usually associated with GRC at a very early age. A similar effect can be observed if the length of the

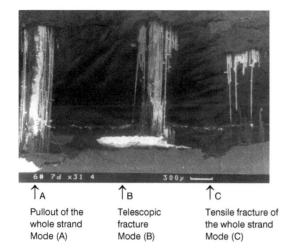

Figure 5.2 A composite of microphotographs of the three typical modes of tensile fracture of strands of glass fibres embedded in a cementitious matrix and crossing a crack in the GRC. The fractures were observed during tensile tests carried out with a scanning electron microscope (SEM).

↑A — Pullout of the whole strand Mode (A)

↑B — Telescopic fracture Mode (B)

↑C — Tensile fracture of the whole strand Mode (C)

strands is very short and the embedment is insufficient to transfer enough stress to break the fibres.

2. *'Telescopic' or 'combined' mode of failure*

In this mode, peripheral fibres of the strand, in contact and bonded well to the matrix, take up most of the additional stress transferred from the cracked matrix and eventually fail in tension.

Their failure then transfers all the stress previously carried by the peripheral fibres onto the next layer of inner fibres within the strand. The inner fibres then begin to fail in stages, and the process continues until all the fibres are broken.

The pullout mode of fracture prevailed, with some telescopic/combined fractures. The process is usually accompanied by partial internal pullouts of the inner fibres as they break at different points, and the frictional part of the bond continues to provide additional means for fibre–fibre transfer of some of the internal stress.

This is the mode of failure that requires the highest amount of work of fracture (the highest pseudo-ductility or toughness is observed), while the

Figure 5.3 Flexural test fracture surface – premix GRC. Almost all of the fibres remained bundled in strands.

strength of the composite is also enhanced. Unfortunately, such a mode of fracture is usually a transient phenomenon, associated generally with GRC at a lower age, in which the magnitudes of external bond (with the matrix) and bond within a strand are in the right proportion. When and for how long this mode of failure exists depends very much on factors such as external service conditions, mix design and age. The failure of the GRC in this mode is gradual, and a complete fracture occurs only after a substantial deformation/extension of the composite. It is the mode of failure most suitable for GRC in practice.

3. *Complete tensile fracture of the strand.*
   This mode occurs when bond both between the fibres within the strand and between the strand and the matrix is very high. As in Mode 1, the strand then acts as a single reinforcing element and fails in a complete tensile fracture with a minimum extent of pulling out, or no pullout at all.

The strength of the composite is enhanced; however, the work of fracture is significantly lower than in Mode 2. The composite shows a typically brittle fracture with a sudden failure at a much lower ultimate strain/deformation than in Mode 2.

A pseudo-ductile composite with failure modes (1) or (2) is much preferred to a brittle one (Mode 3) in structural applications. Fractures in Mode 3 are typical of a highly aged GRC, particularly in a humid environment.

In practice, there are no sharp or distinct boundaries between the three basic modes of fracture mentioned above. More than one type of fracture mode can be present in a given GRC at any age. Moreover, the fracture mechanism is also influenced by factors other than bond, such as the position of a reinforcing strand in relation to the applied stress and the line (or path) of the crack in the matrix.

The fractures shown in Figure 5.2 are typical of a strand bridging a crack in a position perpendicular (at 90°) to the fractured surface of the matrix. In this case, the modes of fracture can be identified. However, in practice, only a very small percentage (5–8%) of strands would be in such a position.

The fracture process of strands crossing a crack at a more acute angle is even more complicated. There are indications [45] that fibres/strands at a slight angle to the principal stress provide more 'toughness' than those perfectly perpendicular, which make up only a small proportion of all strands randomly distributed in a GRC matrix. Strands lying entirely parallel to the applied stress and weakly bonded to the matrix may even reduce the strength of the composite. Limited studies have indicated [46] that the peripheral fibres of the strand fracture first, as the edge of the matrix along a crack concentrates the stress. The broken fibres then provide a 'cushion' over which the inner fibres can still pull out at an angle (Figure 5.4).

The contribution to the fracture process by strands at higher angles to the applied stress is extremely difficult to examine directly. Numerical modelling of the fracture processes in GRC has been tried. However, without access to values for bond obtained independently (experimentally) for all its different aspects and considering its inherently variable nature, no reliable and genuinely predictive numerical model for GRC has been developed to date.

Several important parameters and features are identified, and data very useful for GRC in practice are obtained, from load-deflection or stress-strain diagrams such as those shown in Figure 5.5. LOP (in megapascals) is the stress corresponding to the end of the initially straight line of the diagram. The straight line covers the first stage in the test, in which the GRC responds to the applied load in a completely elastic manner and behaves as a homogeneous composite. The angle of the straight line is directly related to the (Young's) modulus of elasticity E (in gigapascals), which also determines the strain, ε_{LOP}.

The assessment of performance in the early stages of the development of GRC sometimes included a value of the stress at a bend over point (BOP) (in megapascals). The BOP indicated the level of flexural/tensile stress at which the stress-strain or load-deflection curve plotted in a test started to deviate from a straight line and showed a 'bend'. It was possible to associate this with the beginning of a visible and substantial multiple cracking of the

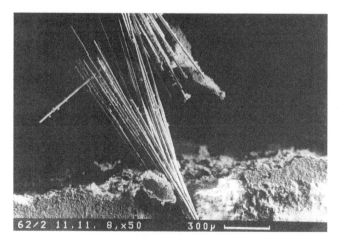

Figure 5.4 Fracture of a strand of fibres crossing a crack at an angle during a test within a SEM. Note the broken fibres at the sides, where the edges of the matrix 'cut' into the fibres. A small number of fibres in the interior of the strand still cross the crack and carry load [46].

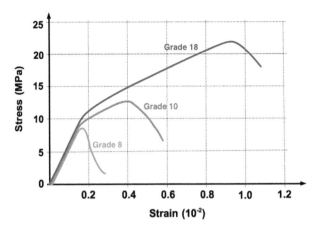

Figure 5.5 Typical shapes of load-deflection (stress-strain) diagrams of GRC in bending [47].

matrix. However, load/extension or stress/strain curves from tests on modern formulations of GRC often make it very difficult to identify the BOP. The LOP is a more important parameter, which is also easier to determine accurately, and it is used instead.

Another important parameter, which has survived from the earliest days of GRC development and which is sometimes difficult to understand in a modern context of flexural behaviour of fibre-reinforced composites in general, is the modulus of rupture (MOR) (in megapascals). The MOR indicates the maximum flexural stress a test specimen was able to resist. It is equivalent to the *bending strength of the material* tested. It is not a 'modulus' such as the modulus of elasticity. Nor is it usually associated with a 'rupture'. As is typical for a pseudo-ductile composite based on a brittle matrix reinforced with fibres, the GRC normally does not fail entirely at the MOR level of stress. Instead, it continues to resist, gradually reducing flexural stresses until the ultimate tensile strain ε_{ult} (%) (deflection, extension) is reached.

Only then does the composite lose its integrity, and its load-carrying capacity is reduced to zero. With the ε_{ult} being normally much greater than the ε_{MOR}, the failure of GRC is usually a gradual process, providing a useful safety margin in its structural applications. A genuine tensile strength (UTS) of the GRC can be obtained only from a *direct uniaxial tensile test.* Such a test produces the value of the ultimate tensile strength as the stress at the peak of the stress-strain curve. As in the bending test, the ultimate tensile strain does not normally coincide with the ultimate tensile stress, but occurs only after an additional strain develops and the tensile stress has already decreased.

5.3 Influencing factors

Principal factors influencing the properties of GRC are outlined below. Considering individual factors, *it is essential to understand that factors which influence the mechanical properties of the GRC are both numerous and inter-related to a very large and variable extent.* The magnitude of the effect of one single factor on a given property usually depends on the value(s) of other factor(s). Such multi-factorial interactions also tend to be non-linear and of a varied degree of significance.

Reliable, general conclusions therefore cannot be drawn from a test series or from an experiment in which only two factors/variables are correlated without regard to any of the others.

5.3.1 Mechanical properties of fibres

The *modulus of elasticity of fibres* (E_f) has a fundamental influence on the properties of the composite.

The effect of E_f depends on its comparison with the elastic modulus of the matrix (E_{matrix}) = approximately 35 GPa (a typical value for hardened cement paste/concrete of the type used in GRC).

Fibres with $E_f > E_{matrix}$ can increase both the toughness and the strength of the composite (typically steel fibres, carbon fibres, etc.). Alkali-resistant glass fibres for GRC with E_f = 72–74 GPa are at the lower end of the range.

Fibres with $E_f < E_{matrix}$ can increase the toughness but cannot increase the strength of the composite. Their addition may decrease the strength of the composite (compressive, tensile), which can be lower than that of the fibre-free matrix alone. A typical case is ordinary concrete reinforced with polypropylene (E_f of approximately 3.5–7.0 GPa) or other relatively low-modulus fibres.

The tensile strength of fibres governs the ultimate direct tensile and flexural strength of the composite. A very high strength can be exploited only if the bond between the fibre and the matrix (and between the fibres within the strand) is high enough and the fibre is long enough to enable the high strength to be fully utilised. If the bond is low, the strand will pull out, carrying load (stress) below its ultimate tensile capability.

The tensile strength of the fibres within the GRC is lower when the cross section of the fibres is reduced by corrosion due to a reaction between the glass and the surrounding highly alkaline cementitious environment. The alkali attack creates localised weak spots on the fibres, which act as defects and substantially reduce the very high tensile strength of the initially pristine fibres. Tensile strength is also reduced when the fibres remain in bundles and crystalline products of hydration of the surrounding cementitious matrix precipitate on the surfaces of individual fibres in the inter-fibre spaces within a bundle. Such products are hard enough to cause local damage to the surface of the fibres and act as stress concentrators, reducing the tensile strength of the fibre; this is when the first fibre–matrix or fibre–fibre displacements occur.

Both the individual fibres and bundles of fibres are flexible. However, when they carry a load and bridge a crack in the matrix at an angle other than 90°, the 'edge' of the crack acts as a severe stress-concentrator, and their 'tensile' load-bearing capacity is also reduced.

5.3.2 Mechanical properties of the matrix

The matrix makes up around 95% of the volume of GRC. The mechanical properties of the matrix largely determine those of the GRC overall, as the basic properties of GRC also follow the 'rule of mixture', which is generally applicable to brittle-matrix composites.

- The *tensile strength and the ultimate tensile strain* of the matrix are key factors controlling the magnitude of the stress at the LOP. The value of LOP enters the basic structural design of GRC elements. Any increase in LOP is therefore directly beneficial to their performance. However, high-strength matrices also tend to have lower ultimate tensile strains, and as bond is also affected, a complex process of optimisation of the matrix properties is required to achieve the best performance of the composite.

- The *modulus of elasticity of the matrix* controls the overall modulus of elasticity of the GRC. In turn, the modulus of the matrix (E_m) depends on the modulus of elasticity of the aggregate and of the hardened cement paste. Most of the GRC matrices are based on standard silica sand, and any significant change of the E_m is therefore achieved by modification of the cement paste. However, raising the strength and therefore the modulus of the hardened cement paste normally leads to an undesirable lowering of its ultimate strain and an undesirable embrittlement of the composite. It is therefore more common to lower the modulus of elasticity of the matrix by addition of a high proportion (10–15% b.w. of cement) of a polymer (usually an acrylic resin) compatible with the cementitious matrix. The GRC then becomes inherently, and proportionally, more ductile, but the polymer is also likely to affect the internal mode of fracture of the glass fibre strands and of the GRC overall.

- *Abrasion resistance*
  There is no specific guidance for the abrasion resistance of GRC. In cases where the abrasion resistance is of significance, an approach similar to that for abrasion-resistant concrete may be adopted (abrasion-resistant aggregate is used), although there is much less scope to modify the GRC matrix to make it more abrasion resistant (lower aggregate content).

- *Hydraulic resistance*
  Smooth mould-side surfaces of GRC make it suitable for elements used in a variety of hydraulic structures. Manning's roughness coefficient, used in the Manning formula for calculation of a hydraulic flow in an open channel, is approximately 0.012 [48].

5.3.3 Bond

The reinforcing effect of fibres depends very strongly on the *fibre–matrix bond*, which, in turn, depends on the micro/nano-properties of the contact zone (often identified as the interfacial transition zone – ITZ) either between a fibre and the matrix or between fibres within a strand.

Two types of bond that control the 'composite' action between matrices and reinforcement can be broadly identified [44].

- *Adhesive bond* tends to be low between glass and hardened cement. It keeps deformations of both the fibre and the matrix proportional/compatible in the early stages of loading. The first crack of the matrix, which occurs when its tensile strain capacity is exceeded, causes an additional load to be transferred to the strand, and an additional interfacial stress is generated. When the interfacial stress exceeds that of the adhesive bond, an adhesive de-bonding begins and proceeds rapidly along the remaining embedded length of the strand.

Following the failure of the adhesive bond, any remaining resistance against pullout of a fibre crossing a crack is largely controlled by frictional bond and any mechanical interlock/anchorage of the fibre in the matrix.

- *Frictional bond* resists further displacement (pullout) of the strand (or fibres within a strand) during multiple cracking of the matrix up to and beyond the ultimate tensile or flexural strength (MOR) of the composite. Frictional bond is neither uniform nor constant; its magnitude changes with the speed of the loading and displacement and with the fracture process.

Both adhesive and frictional bond develop with age (see Section 5.2), the development being strongly influenced by the service environment, by chemical interactions between the matrix and the fibres, and by crystallisation of any products of hydration of cement in the inter-fibre spaces within a strand.

Low adhesive bond means that strands (fibres) perpendicular to the direction of the applied load (stress) not only do not contribute to the strength of the composite, but they may effectively act as 'voids' and have a negative influence. If the strands within GRC are considered as single reinforcing elements, the fibres contributing most to the strength of the composite are likely to be those slightly at an angle to the direction of the applied stress [45].

Glass fibres in GRC are not provided with any 'anchorage' at their ends, unlike many types of fibres in steelfibre reinforced concrete (SFRC). This type of 'mechanical interlock/bond' therefore has a very small, often negligible effect on the GRC fracture processes.

An accurate quantitative assessment of bond within the GRC and its eventual control (design) remain the main and very significant challenge on the path to a greater exploitation of the potentially very high performance of GRC. Direct measurement of the bond by 'pullout' tests, such as those used for ordinary concrete and most other fibre-reinforced concretes, is not practicable in GRC research, where the basic reinforcement is by bundles of fibres.

A breakthrough came with the first applications of nanotechnology in the mid-1990s [49]. A nano-indenter was developed, which permitted precise loads (millinewtons) to be applied onto microscopic targets (10^{-6} m) and which was capable of measuring displacements continuously at nano scale (10^{-9} m). Such extremely high-precision testing can only be carried out in specialist laboratory premises with very closely controlled temperature and in a vibration-free environment. The nano-indentation technique opened up a new approach to investigation of the bond in GRC. It also enabled the 'micromechanical' properties of the matrix, both near the fibre–matrix interfaces and in the bulk of the matrix or the fibre, to be assessed.

A single-fibre push-out test for testing of bond in GRC was developed. A very thin (<0.5 mm) slice was cut perpendicular to the axial direction of a strand of glass fibres embedded in the matrix material to produce a push-through test specimen. The specimen was placed on a special supporting plate with a small opening and moved to a position in which only a very small, central part of a chosen strand would remain unsupported and permit a push-through. The whole set-up was put into the nano-indenter and manipulated to a position in which the tip of the indenter was above the centre of the selected fibre. Backlighting from below the specimen lit up individual fibres and facilitated the positioning of the indenter. A successful test produced a diagram of load (millinewtons) against displacement (nanometres).

The original instrument applied the load via a pointed diamond indenter, which was the easiest to produce, using established techniques for shaping diamonds. However, the pointed tip also made an indentation into the single glass fibre tested. It provided useful additional information about the mechanical properties of the glass in the fibre itself (load displacement when the glass was indented); however, it often caused splitting of the fibre (Figure 5.6), which invalidated the bond test results. The indenter was later modified by reshaping the tip of the diamond used in order to provide a flat contact area with the cross section of the fibre tested [50]. Reshaping of the diamond tips to the form of a truncated pyramid with a flat tip approximately 5–7 μm wide eliminated the splitting of the fibres. However, diamonds cannot be cut into shapes of truncated cones or pyramids by traditional methods. Such shaping was only possible by using focused ion beam technology, a very advanced technique, which permits 'machining' even hard materials such as diamonds with great precision into flat-tip indenters. The cost of production of such indenters was very high, and this curtailed further progress in a systematic investigation of bond.

The first direct information about the bond both between fibres within a strand and between peripheral fibres and the surrounding matrix was obtained from the project described in this section [50], but much more work is still required to build up an adequate database on all aspects of bond. Such data will eventually permit adequate control of the bond within GRC. This will help to make a substantial advance towards full exploitation of the structural potential of GRC.

The basic influences of bond in its different forms are now understood; however, their relationship with the mechanical properties of hardened GRC still remains inadequately explored.

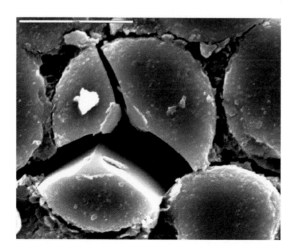

Figure 5.6 Cross section of a single glass fibre within a strand, split by a sharp diamond indenter in a nano-indentation 'push-through' test [50].

The absence of an adequate database of reliable quantitative information on bond, obtained from independent tests, makes it still largely impossible to develop genuinely predictive numerical models of fracture of GRC. The numerical models proposed either apply only to a certain stage in the fracture process or are only 'descriptive'. Descriptive models provide numerical simulations of the results of a given test with any bond characteristics adjusted to make the description fit the results. Such models cannot reliably predict the performance of GRC.

Properties of GRC | 6

6.1   Properties of fresh GRC

The workability of the fresh matrix material must be appropriate for the production method selected. Generally, the higher the workability, the easier the production and the compaction. However, high workability must not compromise the expected high strength of hardened GRC or produce a fresh mix susceptible to an unacceptable degree of bleeding. The mix design of common GRC matrices therefore includes superplasticisers, which permit the production of highly workable mixes without a reduction of strength. Admixtures that have been specifically developed for self-compacting concrete are also used in self-compacting premix GRC.

Workability has to be maintained within narrow limits and requires to be monitored. Workability is usually measured on freshly mixed slurry (matrix) before the addition of fibres. A standardised mini-slump test [51] is available for this purpose. The test serves both as an indicator of workability and as an indirect measure of uniformity of composition of the fresh GRC produced; it fulfils the same function as that of the slump test in standard concrete technology.

The basic arrangement of the test and the equipment is shown in Figures 6.1 and 6.2. A cylindrical mould (internal diameter 57 mm, length 55 mm) made of a transparent polymer (Perspex/Plexiglas) is placed in the centre of a plate with engraved concentric rings. The rings vary from 65 to 225 mm in diameter, numbered from 0 to 8. The mould is filled with the freshly mixed slurry in the required manner and then raised vertically. The number of the ring touched by the slurry once it stops flowing is recorded.

The test is carried out three times, and the average result is calculated, rounding to the nearest ½ ring number. It is important to record the time

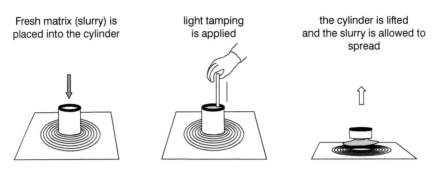

| Fresh matrix (slurry) is placed into the cylinder | light tamping is applied | the cylinder is lifted and the slurry is allowed to spread |

Figure 6.1 Basic arrangement of the (mini) slump test for fresh slurry [51].

K&C MOULDINGS LTD

Figure 6.2 Baseplate and mould for the mini-slump test.

elapsed between the end of mixing and the mini-slump test being done, and the ambient temperature. Spreads between ring numbers 2 and 5 usually indicate that the slurry (fresh matrix) is suitable for production of a fresh GRC by spraying.

The test helps in production control to ensure that the optimum workability required for the chosen method of production is maintained. The test is described in more detail in Section A.3.

Freshly produced GRC in the form of 'green' sheets can be re-shaped by folding and pressing to produce 3-D elements. Such sheets can be made by spraying or extrusion or premix process, reinforced with continuous strands of alkali-resistant (AR) glass fibre or woven or non-woven meshes.

Care must be taken to ensure that fibres are not damaged in the process and that no cracking or other discontinuity is generated in corners and bends. This method would generally use moulds with folding sides, and it is particularly suitable for making largely flat, tray-shaped products.

The glass content of freshly produced, uncured, green GRC can be verified by using a 'wash-out' test, described in detail in Appendix B.

6.2 Properties of hardened GRC

6.2.1 Basic influencing factors and typical mix designs

The properties of hardened GRC depend on many factors, which are often strongly and to a variable degree inter-dependent. These can be grouped. They include constituent materials (described in Chapter 3) such as

- Matrix (type of cement, water content to cement content ratio [w/c], cement content, additives and admixtures);
- Filaments (size/diameter, strength, modulus of elasticity, surface treatment);
- Strands (number of filaments, shape, length).

Their proportions are established in the process of mix design, which aims to produce hardened GRC with the performance characteristics required.

The mix design takes into consideration the effects of additional factors, such as

- Production process, which includes mixing process, compaction, curing, distribution and orientation of fibres/strands;
- Application of the product, which determines the length of service life (age) and expected exposure/service environment (predicted levels/variations of temperature and humidity, including freeze/thaw cycles), chemical attack, biological attack and any chemical/physical fibre–matrix interaction;
- Development/changes in fibre–matrix interaction over the period of service life (age) and conditions for a given GRC mix design.

The data in Table 6.1 provide an aid for the selection of grades of GRC for a range of applications. Ultimately, the choice of grades has to be confirmed by the specifier, a specialist GRC manufacturer and a competent engineer. The guide in Table 6.1 is based on UK experience and may vary for different countries.

Mix proportions for typical grades of GRC, both with and without the addition of a polymer for curing, are provided in Table 6.2 [32]. The mix designs in Table 6.2 are also intended as a guide. Designs with proportions falling outside these guidelines may still be acceptable, but the performance of such GRC would have to be verified by trials before use.

Table 6.1 Guide to selection of grades of GRC for typical application. (After Ian White)

Market sector	Typical application	Component size/face area	GRC production process	Grade (MOR)	Curing method		Testing frequency
					Polymer	Moist	
Architectural	Claddings, Soffits, Column encasement	1 m² or over	Sprayable	18	Yes		Daily
	Architectural features	Under 1 m²	Pourable cast or sprayable premix	10	Yes		Weekly
	Built-in components, heads, cills, band courses	Self-supporting, Non-load-bearing	Pourable cast, Pourable cast or sprayable premix	8/10, 8/10	Yes, Yes		Weekly, Weekly
	Architectural perforated sunscreens	2 m² or over, Under 2 m²	Discuss with manufacturer, Pourable cast or sprayable premix	10	Yes, Yes		Weekly, Weekly
Civil engineering	Permanent formwork	All	Sprayable	18		Yes	Daily
	Drainage – general		Pourable cast or sprayable premix	10		Yes	Weekly
	Drainage – large items, Retaining structures			18		Yes	Daily
Other products	Plant pots, Street furniture, Garden ornaments	–	Pourable cast or sprayable premix	8		Yes	Weekly

Table 6.2 Premix grades 8, 8P, 10 and 10P, and sprayed grades 18 and 18P

Premix grade	Grade 8	Grade 8P	Grade 10	Grade 10P	Grade 18	Grade 18P
Process	General-purpose premix		Sprayed premix or high-quality cast premix		Sprayed GRC	
Aggregate/cement ratio	0.5–1.50	0.5–1.50	0.5–1.50	0.5–1.50	0.5–1.5	0.5–1.5
Water/cement ratio	0.30–0.40	0.30–0.40	0.30–0.38	0.30–0.38	0.30–0.38	0.30–0.38
Glass fibre content (% b.w. of total mix)	2.0–3.0	2.0–3.0	2.0–3.5	2.0–3.5	4.0–5.5	4.0–5.5
Polymer solids content (% b.w. of cement)	Nil	4–7	Nil	4–7	Nil	4–7

6.2.2 Methods for assessment of performance

The basic ingredient of GRC is a cement-based matrix, the properties of which develop with age. Age is therefore one of the key factors influencing the properties of hardened GRC. However, it is impractical to assess its effect directly in a systematic manner because of the relatively long service lives of GRC structures/products. Accelerated ageing tests are therefore used, whereby the effects of a long natural exposure are replaced by a shorter exposure in more aggressive warm/hot solutions simulating pore solutions in a specific matrix material. Accelerated ageing tests are useful for comparative purposes; however, performance in real service life may depend on additional factors, such as the history and effects of applied stresses and micro- or macro-cracking already present in the composite. A specific type of accelerated ageing test may not be appropriate for all types of mix design of GRC.

Most of the properties of GRC that change with age and conditions/ exposure in service reflect the progressive hydration of cement in the matrix. Hydration of cement continues for a long time, albeit at a decreasing rate. Simultaneously, changes in fibre–matrix interaction (chemical and mechanical, bond) occur during ageing and at a non-linear rate.

It is therefore essential to quote the age and curing/exposure history whenever properties of hardened GRC are mentioned or comparisons made.

Cementitious matrix itself is inherently a highly variable material, and the same therefore applies to GRC overall. Any reliable quantification of its properties, therefore, has to be supported by an appropriate statistical measure of this variability. This is already a well-established approach, and most of the properties of hardened GRC are quoted as 'characteristic' values. Values that have a 95% probability of being matched or exceeded by the GRC tested are normally used as a default value, unless (rarely) another probability level is selected.

Average values of a property therefore carry only a 50% probability. Average values must be specifically indicated and preferably accompanied by the size of the population (number of tests) from which the average had been obtained. This may also permit the characteristic value to be determined.

The dispersion of fibres within GRC is never completely random or uniform. The composite, therefore, always exhibits a variable degree of anisotropy, which means that its performance parameters also depend on the direction of the applied load. This is particularly significant for all types of strength-related properties, where the value of a property depends on the angle between the direction of the applied load and that of the prevailing number of fibres.

A large proportion of GRC is produced as flat sheets in which the length of fibres (12–37 mm) exceeds the thickness (10–20 mm) of the sheet. In such cases, the fibres tend to align automatically with the plane of the formwork, generating a distinctly anisotropic composite. However, such anisotropy can be exploited and even deliberately enhanced by the incorporation of two- or even three-directional meshes made from continuous strands placed in the most effective positions and orientation to improve the performance of a GRC element.

6.3 Physical properties

6.3.1 Density

The density (bulk density in kilograms per cubic metre) of hardened GRC depends primarily on the density of the matrix, which, in turn, depends on the content and density of the aggregate and on the degree of compaction of the fresh composite. The free water content and any trapped/entrained air of the hardened GRC also influence the density. Specimens of the composite for assessment of density and exact/reliable comparisons should be oven-dry, or of identical moisture content.

A range of densities of hardened, dry, 'ordinary' GRC is shown in Table 6.3.

As is the case in standard concrete technology, replacing an ordinary aggregate with a lightweight one reduces the overall density but with an inevitable and significant reduction in strength. In relatively rare applications, where such a reduction of strength can be tolerated, the density of GRC can be reduced to approximately 1600 kg/m³.

The grading of the lightweight aggregate must still comply with the basic requirement of maximum particle size appropriate for the selected GRC production method. Preferred types of lightweight aggregate, such as perlite, consist of strong spherical particles with closed surfaces, which makes it easier to control the water content of the mix.

It should be noted that the high density of standard GRC is generally considered an indirect indicator of good quality. Test results with values near to that of a

Table 6.3 Densities of typical GRC mixes

	Spray grade		Premix/sprayed premix grade	
	Grade 18	Grade 18P	Grade 10	Grade 10P
Minimum bulk dry density (kg/m³)	1800	1800	1800	1800
Minimum bulk wet density (kg/m³)	2000	2000	2000	2000

theoretical density calculated from the specific GRC mix design show a very good compaction and are likely to achieve the predicted properties when hardened.

6.3.2 Permeability, water absorption and apparent porosity

A well-compacted GRC has an inherently *low water permeability*, and it is considered a waterproof material, suitable for use in water-retaining structures. Specific performance characteristics are strongly dependent on the mix design and production method of GRC.

Permeance by water, measured on samples of 8 mm thick freshly made GRC, is in the range of 0.02–0.4 ml/m²/min [52].

Water absorption (5–11%) and *apparent porosity* (16–25%) are similar for both the premix and sprayed GRC [32, 52].

Standardised tests, for example EN 1170 Pt. 6 [53] or the Glassfibre Reinforced Concrete Association (GRCA) test method [54], are used for assessment of the absorption, based on measurement of the mass of GRC in an oven-dry condition (dried to a constant weight) and a saturated condition (immersed in water until a constant weight) and the weight of the sample when suspended in water.

6.3.3 Acoustic properties

The high density of a typical GRC provides it with an inherently good capacity for *attenuation of noise*. However, the 'mass law', which indicates how the rate of sound transmission reduces when passing through denser materials (see Figure 7.6), becomes ineffective for sound frequencies near to the 'critical frequency' of the whole element (panel), including its fixings. The critical frequency occurs when the wavelengths of the sound and of the structural (flexing) response of the panel match each other. The acoustic performance, therefore, depends not only on the intrinsic capacity for noise attenuation of the material used but also on the geometry and method of fixing of the GRC element.

A sheet of GRC 10 mm thick, with a density of 2000 kg/m³, subjected to the sound pressure of 0.2 kPa, will provide a Sound Transmission Class of 34 [55, 56]. The increase in attenuation (reduction) of sound follows well the mass law (Figure 6.3) when the frequency of sound is below one-half of the critical value for a given test specimen (panel).

The attenuation of sound is a well-recognised property of GRC, and the composite is widely used for the construction of sound barriers alongside busy roads/motorways and railway lines, among other applications [57]. GRC sound barriers do not rely on the 'mass law' alone (a 10 mm thick GRC panel will provide 20 kg of mass per square metre); they are also designed to cause diffraction of the sound.

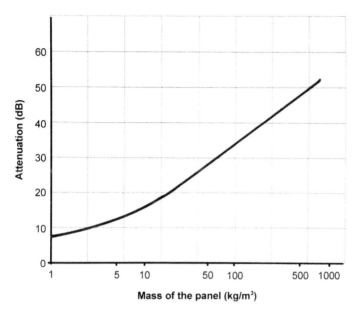

Figure 6.3 Reduction in transmission of sound in relation to mass of a panel [56].

GRC panels as sound barriers are designed either to produce a surface profiled to such an extent that sound is reflected and interferes with itself or to reduce the transmission of sound by absorption. Multifunctional panels can be produced using a 'sandwich' construction. Sandwich panels can offer sound reflection or refraction on one side and absorption on the other.

6.3.4 Thermal properties

The thermal expansion/contraction of GRC is described by a coefficient indicating 'unit strain' for one degree of change in temperature. The coefficient has a range of approximately 10–$20 \times 10^{-6}/°C$. Exposure conditions have an effect; minimum values of expansion/contraction occur at high and low relative humidity (RH). The maximum value occurs at around 50–80% RH.

Dimensional changes due to the variation of temperature have to be considered along with moisture and shrinkage movements when designing GRC products.

- *Thermal conductivity and thermal resistance* depend on the density of the GRC and its moisture content.
- *Thermal conductivity* (λ) of a typical GRC with density of 1900–2100 kg/m^3 is in the range of 0.5–1.0 W/m°C. It is a characteristic of the material.

- *Thermal resistance* (R) depends on the thickness (t) of the material and its thermal conductivity (λ); $R = t/\lambda$.

- *Insulating capability of a building element* U is an inverse of the resistance R plus the heat losses by convection and radiation. The value of U, therefore, depends on external factors reflecting the environmental situation of the building element in addition to the thermal resistance of the material.

6.4 Mechanical properties

Typical values of current basic mechanical properties at the age of 28 days are shown in Table 6.4. Routine testing for verification of the mechanical properties of GRC concentrates on standardised flexural/bending tests [58].

Direct tensile tests provide the best characterisation of the fracture process and strength of GRC, but they are expensive and limited to specialist testing laboratories with advanced equipment. It is inherently difficult to ensure a genuine uniaxial loading of the test specimens throughout a test and prevent failures within the area of end grips. The latter can be reduced by shaping of the test specimens (enlarged ends, such as a 'dog-bone' shape), but the production of such test specimens increases the costs.

Practical experience over many decades has shown that correctly set-up and performed *flexural tests* provide a good indication of the strength of GRC and its variability for use in practical design and applications.

6.4.1 Modulus of elasticity

Young's modulus of GRC reflects primarily that of the matrix, with the modulus of elasticity of fibres making a contribution. The latter has a less direct effect, as it depends on the degree of stress transfer available, which is controlled by the fibre–matrix bond.

In the *'elastic' region of fracture* process, before the stress reaches the LOP, the modulus of elasticity Ec of the GRC can be estimated from a combination of E_f and E_m using the law of 'mixtures', as is applied to brittle-matrix composite materials in general.

The elasticity of the composite can therefore be adjusted beyond the 'normal' range given in Table 6.1 by a significant modification of the modulus of elasticity of the matrix. The adjustment is more often downward, to make the hardened composite more 'ductile'. In such case, the ultimate tensile strain and the toughness of the composite are increased. On the contrary, a very high modulus (and usually also a very high strength) of the matrix tends to make the composite more brittle.

Table 6.4 Range of properties achieved by a typical hardened premix or sprayed GRC [54]

| Dry density (kg/m³) | Ultimate strength (MPa) at 28 days in | | | Modulus of elasticity (GPa) | Tensile strain at MOR (%) | Impact resistance (kJ/m²) | Thermal movement | Thermal movement | Poisson's ratio |
	Bending (MOR)	Uniaxial tension	Transverse tension						
1900–2100; *1800–2000*	5–14; *18–30*	3–6; *8–12*	7–11; *5–8*	10–20 (both)	0.1–*0.8*	7–12; *15–25*	$10\text{–}20 \times 10^{-6}$ per °C (both)	$300\text{–}1200 \times 10^{-6}$ (both)	0.24–0.25 (both)

Note: premix values are in italics.

6.4.2 Flexural/bending strength

A load-deflection or stress-strain diagram of GRC tested in bending is a key performance parameter of the composite. It provides fundamental information for a safe structural design and indicates the fracture mechanisms involved. Figure 6.4 shows typical shapes of simplified load/extension diagrams. Bending (flexural) strength is the principal parameter of hardened GRC. It is assessed by a standard test [58], which can be carried out in two versions, depending on the sophistication of the test equipment. A typical test arrangement is shown in Figure 6.4 and typical test equipment in Figure 6.5. The process is described in Appendix C.

Values of MOR and ultimate strain decline non-uniformly with age and exposure conditions, eventually stabilising at a level a little greater than the value of the long-term LOP. As the hydration of the cement in the matrix continues, the LOP itself tends to rise, and the MOR reduces slowly with age (Figure 6.6).

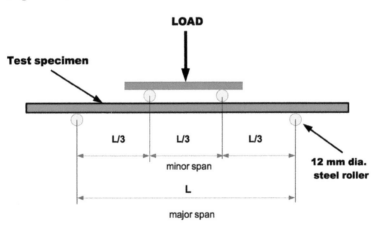

Figure 6.4 Layout of the standard flexural (bending) test on GRC.

Figure 6.5 Typical test jig for the standard bending test.

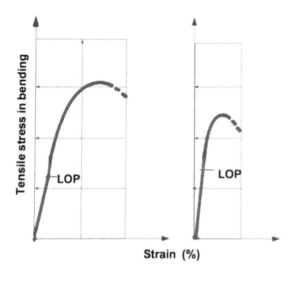

Figure 6.6 Development of LOP and MOR with age [59].

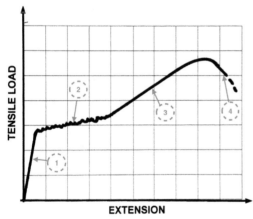

Figure 6.7 Typical stress-strain diagram for GRC in tension.

6.4.3 Tensile strength and Poisson's ratio

The behaviour of a GRC composite is best characterised by its load-deformation (or stress-strain) diagram, obtained from a direct uniaxial tensile test (Figure 6.7). However, this type of test is difficult to carry out reliably by simple test equipment (difficulties with gripping of specimens, size-effect, etc.). Testing for direct tension is therefore limited to specialist laboratories.

Note four distinctive stages: (1) elastic region, up to the LOP, when Stage 2 begins; (2) multiple cracking, gradual transfer of the load from the matrix onto the inner fibres of the fibre strands; (3) matrix is heavily cracked; with adequate bond, fibre strands can carry increasing load, peaking at the value of MOR; (4) fibres gradually fail and load declines down to zero, when the test specimen loses all integrity.

A much simpler four-point bending test is therefore carried out in GRC practice (Figure 6.4). Tensile strength is then calculated from the results on the basis of an assumed elastic stress distribution occurring during the standard bending test. A range of ultimate tensile strengths is given in Table 6.4 for sprayed and premix GRC when tested parallel to the orientation of fibres at 28 days.

The anisotropy of GRC has an influence on the direct tensile strength. With almost all fibres aligned to the direction of the tensile stress, the ultimate tensile load (up to the first crack) may be higher than that of an unreinforced matrix alone. However, a very high bond both between the matrix and the fibres and between fibres in a strand must exist for such an enhancement of tensile strength to occur.

Poisson's ratio describes the ratio between strain parallel and strain perpendicular to the applied tensile stress. It can be measured only as part of a direct tensile test [8], which is not carried out routinely. The Poisson's ratio of a typical GRC is approximately 0.24.

6.4.4 Compressive strength

A typical GRC load-bearing element is made of relatively thin sheets of the composite. In such a case, the compressive strength alone tends to be only a secondary parameter, the critical property being the flexural (bending) or tensile strength. Applications in which bulk GRC is used are usually nonstructural, so there is no need for a standard procedure to test GRC for its compressive strength.

The small amount of data available on compressive strength is difficult to compare. Testing in compression is complicated by additional variations in results caused by the degree of anisotropy of the composite in a specific test specimen, even if a suitably bulky 3-D GRC test specimen was produced. The compressive strength of GRC generally reflects the compressive strength of the matrix alone, which can be enhanced or reduced by the presence of fibres, depending on their orientation, content and bond. In cases where the compressive strength is of importance, tests are more likely to be carried out on prototype elements, subjected to loads expected in service, instead of individual generic test specimens. It would often be very difficult to obtain test specimens that would reliably represent the GRC in a specific application/product to be assessed.

For approximate guidance only, the compressive strength of GRC can be estimated as 40–60 or 50–80 MPa for cast premix and spray-up process, respectively [52].

6.4.5 Transverse tensile and inter-laminar shear strength

Transverse and inter-laminar shear strength are closely related, both depending largely on the tensile strength of an unreinforced matrix. Simplified loading conditions are shown in Figure 6.8. Any testing is also dependent on the position and geometry of supports.

Transverse strength is measured when a tensile load is applied perpendicular to the surface of a sheet-like GRC element. The strength of the composite in this direction relies greatly on adhesive bond between the matrix and the reinforcement. Adhesive bond between fibres and matrix in a typical GRC is very low. As the position of the fibres is largely perpendicular to the applied load, such fibres may act effectively as 'voids' and cause a reduction of the strength of the matrix in this direction. Transverse strength is of significance for load transfer between rigid fixing elements and a typical GRC thin-sheet element.

There is no standard test for this property as such, any measurement being strongly dependent on the size/shape of any test element and the manner of its support. In practice, it is only tested on full-scale elements or part elements when the performance of a fixing for a specific application is being verified directly. Support conditions in such tests have to be as close as possible to those encountered by the element in service.

Inter-laminar shear strength resists loads applied 'in-plane' of a flat thin-walled GRC element. This is of significance during fracture of the GRC in bending, and it could be of significance in bulky sprayed GRC elements, which may be subject to 'splitting' load at their ends. In the case of bending applied to a flat GRC panel, shear stress parallel to the plane of the panel is generated, transferring stress between surfaces subjected to tension and compression. The inter-laminar shear strength ensures that the element in bending will not fail by an internal displacement, a '*delamination*'. The value of this strength is estimated as 3–5 MPa, which is lower than the tensile strength of the matrix concerned [52]. There is no standard test, and few data are available.

Shear strength and punching shear strength perpendicular to the plane of a typical flat sheet of GRC are inter-related. The shear strength is difficult to

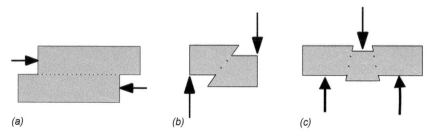

(a)   (b)   (c)

Figure 6.8 Typical shear-loading conditions: (a) inter-laminar shear; (b) in-plane shear; (c) punching shear.

assess as a material's property, as it is strongly dependent on the shape/size and support conditions of any test specimen. In principle, the punching resistance improves with higher fibre content, and then it follows any changes in the ultimate bending strength (MOR) with age. Punching shear strength can be estimated as 25–35 MPa (sprayed) and 4–6 MPa (premix) in typical GRC. Full-scale load trials simulating conditions in service are required for verification of the performance of any structural GRC elements in which punching resistance is considered significant.

6.5   Toughness and impact resistance

The basic constituents of GRC are cementitious matrices and glass fibres; both are inherently brittle materials, and a brittle composite would normally be expected when they were combined.

However, GRC is unusual amongst the very many fibre-reinforced composites in use. The fibre reinforcement in GRC is not by individual, single fibres but by bundles (strands), each containing up to about 200 (original, production number) individual fibres. Bond between the matrix and fibres and between fibres themselves is such that an applied load generates complex stresses and displacements (see Section 5.2), which give the composite a degree of inelastic behaviour; a *pseudo-ductility* is observed. GRC is therefore inherently more impact resistant than plain concrete and many other fibre-reinforced composites. The fracture of a typical mature GRC includes both tensile fractures and pullouts; the highest work of fracture is required when the intermediate/telescopic fractures of the glass strands dominate. It is all strongly related to bond within the GRC and its development with age. Longer fibres provide longer 'critical lengths', the embedded length required for a fibre to break rather than pull out. This explains the higher values of toughness obtained for the sprayed GRC, in which longer fibres are used.

Impact resistance is assessed by standard Charpy or Izod tests using test specimens 25–50 mm wide and 6–12 mm thick, cut from testing boards or finished products. The values shown in Table 6.4 give the range of impact strength for sprayed and premix GRC.

Impact resistance, together with the 'work of fracture' measured as the area below the stress-strain (or load-deflection) curve, provides a measure of the toughness of the composite.

Toughness does not normally enter into current ordinary structural design calculations; however, with all other parameters being equal (bending and tensile strength, etc.), a tougher composite is always preferred. A tough composite usually exhibits substantial deformations before a final break and a general loss of integrity, a useful warning stage before a failure in practical applications.

6.6 Durability

The overall durability of GRC reflects a combination of effects of

1. alkali resistance of the fibres;
2. degree of alkalinity of the matrix;
3. resistance of the cement-based matrix and of the composite overall to degradation caused by exposure to different aspects of the service environment.

Degradation under (3) includes a range of factors, such as wetting/drying and freeze/thaw cycles. In addition, the durability of a GRC product depends not only on the intrinsic properties of the composite but also on the geometry and durability of fixings adopted for a specific GRC element.

In addition to the common factors identified above, durability may also be influenced by the type of coating applied to the fibres during their production.

6.6.1 Wet/dry cycles

Rainfall on GRC causes a rapid wetting, followed by drying, the rate of which depends on the amount of sunshine and the velocity of the air movement. The wet/dry cycling therefore generates volume changes of the hardened GRC. Depending on size, shape, support conditions and any other restraint, *significant internal stresses may be generated within the composite.* Locally, such stresses can exceed the tensile strength of the composite and cause cracking. Such cracking, after many repetitions, can lead to an unacceptable decrease in the overall performance of the product.

Genuine long-term exposure tests to check the wet/dry resistance are impracticable because of the timescale involved. A test in which the effects of wet/dry cycling on the bending strength of GRC are measured has therefore been developed and become standardised [60]. The test simulates the wet/dry cycling in real exposure conditions; however, other factors that are likely to influence the results may have to be taken into account (e.g. the progress and degree of carbonation over a long time) in any particular case. The 'ageing performance' is measured by a ratio (L) of the selected property determined by tests on a reference (not aged) and on 'aged' composite. The reference specimens are tested at the age of 28 days, being immersed in water only for the last day. The ageing process includes the same conditioning for the first 27 days as for the reference material, which is immediately followed by 50 cycles of controlled wetting and drying, each cycle taking 48 hours. One complete ageing process therefore takes 127 days.

The test equipment requires a well-controlled climatic chamber, and it is complex and time-consuming. The test is available only in a very few specialist laboratories, and it is therefore used for research and development alone. It is not practical or economical enough for routine performance testing.

6.6.2 Freeze/thaw cycles

The assessment of frost resistance is based on the same approach as that adopted for concrete. Typical GRC specimens are subjected to cyclic freezing and thawing in a water-saturated state, simulating the worst potential external exposure in service.

Freeze/thaw cycling has an effect similar to that of long-term ageing: the LOP is increased while the MOR decreases, both by up to 20% in extreme cases [53, 56]. The cementitious matrix of the GRC has a very low water/cement ratio, and it therefore performs as well as or better than good frost-resistant concrete, without the need for air entrainment.

6.6.3 Fire resistance

GRC products, when used as cladding, are elements based on thin sections, which heat up much more rapidly than the more massive ordinary concrete elements. The basic constituents of GRC are inorganic, and the basic composite is entirely non-combustible. The fire performance classification according to standard test procedure is *A1* [61, 62].

GRCs with small amounts of organic admixtures (approximately <6% b.w.), such as the acrylic polymer curing compounds, will not sustain burning, but fumes will be emitted. Fire tests classification is *A2*, with a rating of *S1* for smoke production and *d0* for flaming droplet production.

The cementitious matrix will desiccate in fire and undergo both drying shrinkage and thermal expansion. Thermal movement will develop, and when restrained by fixings or the shape of the elements, it will generate excessive tensile stresses and, subsequent cracking. Glass in general softens with temperature starting from approximately 200–700°C, and in the case of AR glass fibres, the tensile strength of the fibres is significantly reduced.

There is relatively little information or evidence about the intrinsic fire resistance of different GRC formulations. In practice, performance in fire is very often dominated by the shape and size of the product. Whenever fire resistance becomes one of the key performance requirements, appropriate standard tests for endurance in fire are therefore carried out on prototype full-scale elements.

6.6.4 Chemical, biological and other exposure

The very high alkalinity of traditional reinforced concrete in contact with air is significantly lowered by interaction with CO_2 in the air, causing 'carbonation'. Depending on the quality of the concrete and environmental conditions, the depth of carbonation can become greater than the thickness of cover to steel reinforcing bars, leaving them susceptible to serious corrosion by oxidation and an accelerating deterioration.

Carbonation of the cement-rich and relatively dense GRC is very much slower than that of good ordinary concrete, reaching only few millimetres after many years in service. With no ferrous reinforcement within the composite and with no effect of the glass fibres, any lowering of the alkalinity of the matrix by carbonation actually benefits the GRC by slightly increasing the strength of the matrix (see Chapter 3).

The effects of acids on the common GRC (ordinary Portland cement based) are similar to those on ordinary concrete, causing an erosion of the cementitious matrix.

There is no evidence of a biological attack of significance reported for GRC in practice. The effects of other chemicals have not been studied specifically for GRC but its resistance is generally comparable to that of a very cement-rich ordinary concrete.

No negative results have been recorded for the performance of GRC subjected to prolonged exposure to UV light. No reduction of the mechanical properties of GRC has been reported when subjected to gamma radiation [52].

6.7 Volume (dimensional) changes

Dimensional changes are of primary significance in the design of GRC elements. Fixing systems must be able to accommodate not only the weight and imposed loads but also all the movement due to changes in humidity and temperature in service. Failure to estimate reliably the maximum potential dimensional changes and to make an adequate allowance will lead to unexpected and often substantial additional stresses being generated both within the GRC element and in the fixings.

In the case of large GRC elements, where shrinkage strain can generate large movements, the mix design may also be modified to produce a 'low-shrinkage' matrix. This is usually done by choosing GRC with the highest practicable content of aggregate (sand) and/or by a modification of the binder. The latter can be achieved by using special, usually blended cements.

The total dimensional change reflects a combination of movements due to changes in humidity and in temperature. It can either reduce or enhance the total movement.

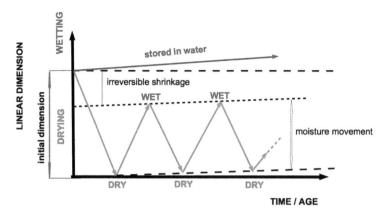

Figure 6.9 Dimensional changes related to humidity of service environment [52].

6.7.1 Effects of humidity: shrinkage/swelling

All materials with a high content of ordinary cement show dimensional changes in response to changes in the RH of their service environment. Figure 6.9 shows that following the initial and irreversible shrinkage due to the progressive hydration of cement, the dimensional changes of the composite reflect changes in the humidity of the environment. The dimensional changes are all non-linear and time dependent. In principle, immersion in water causes expansion/swelling, while drying-out causes contraction/shrinkage; both are reversible, and their magnitudes are time-dependent. Volume changes due to changes in humidity are assessed by a standard test [63].

For design purposes, the initial, irreversible shrinkage ε_{si} can be estimated at 300×10^{-6} with a total long-term shrinkage εtotal of approximately 1200×10^{-6}. Values between 1000 and 1500×10^{-6} are used in the practical design of joints and fixings [52, 59, 63], depending on exposure conditions and the GRC mix design.

As the aggregate used is normally not susceptible to drying shrinkage, increasing the content of aggregate lowers the ultimate shrinkage of the composite. Use of some of the low-alkali-generating cements also reduces drying shrinkage.

6.7.2 Creep and fatigue

As is the case with all cement-based composites, a sustained, long-term loading of GRC produces an *additional deformation (creep strain)* beyond the initial, instantaneous deformation. It follows the same non-linear relationship with age, in which the increase in creep strain diminishes with age. The instantaneous deformation is largely elastic and reversible when the stress applied is less than that of the LOP.

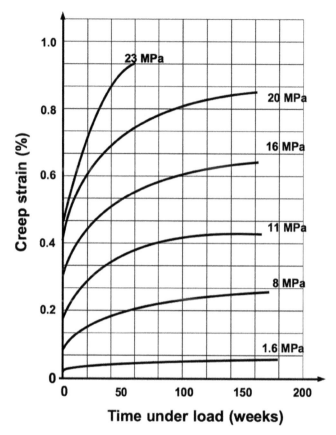

Figure 6.10 Magnitude of the creep strain related to age and level of the applied flexural stress [52].

Sustained loads at higher levels produce greater creep strains (Figure 6.10). Stresses at levels very near to the MOR can cause a delayed, gradual failure of the composite, which is difficult to predict.

Most of the structural design of GRC is based on a permissible maximum load generating a flexural stress less than that of the LOP. In such a case, the creep strain increases at a low rate from the beginning, the 'final' creep strain at high age eventually reaching values between two and four times that of the elastic strain at LOP. Compressive loads are expected to generate levels of creep strain similar to those for flexural stress.

Cyclic and other types of sustained load of variable magnitude affect the performance of GRC and generate fatigue of the composite. Strength is reduced, and the resistance to fatigue is measured by the number of repeats/cycles of the load necessary for a failure to occur, the relationship being plotted as an S-N curve. A typical shape of the S-N curve for GRC is shown in Figure 6.11.

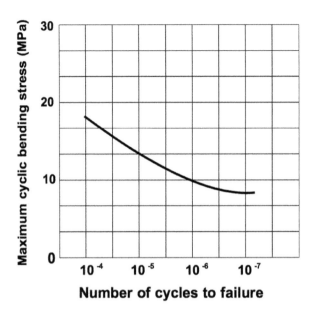

Figure 6.11 Flexural fatigue of GRC (5% of fibres; w/c = 0.33) as a function of peak stress [52].

6.7.3 Thermal contraction/expansion

Mature GRC reflects changes in ambient temperature in the same way as concrete with a very high cement content. Similarly, the coefficient depends on the moisture content of the GRC (matrix); it is lower for very dry and saturated conditions than at 50% RH.

The coefficient of thermal contraction/expansion (αT) is the strain generated by a temperature difference of 1°C. It varies between 10 and 20×10^{-6}/°C, depending on the moisture content of the composite. Maximum values occur for RH of 50–80%; minimum values are associated with an oven-dry or saturated GRC.

It is essential to consider dimensional changes not only due to ambient temperature but also due to the absorption of sunlight on external GRC surfaces, that is, where dark colours were used.

6.8   Self-cleaning

The *e*GRC is a new type of GRC based on the exploitation of *photocatalysis* (Figure 3.5). In addition to its established advantages, such as a reduction in the concentration of pollutants in the surrounding air, it offers surfaces with a significant level of self-cleaning capability. The photocatalytic activity of the *e*GRC surfaces comes from the presence of a nano-crystalline form of the anatase type of TiO_2 in the facing layer. The photocatalytic surface becomes chemically very active in the presence of daylight (or any other source of UV light) and generates a process by which complex organic molecules of common pollutants are broken down by oxidation and removed from the air.

The presence of the photocatalytic TiO_2 also affects the hydrophobic properties of the surface. The UV light changes the original hydrophobic surface (repelling water) into a very highly hydrophilic one (low contact angle; water adheres and spreads). The hydrophilicity of the surface, expressed by the contact angle between the surface and the liquid (water) in contact with the surface, is reduced to less than 5°, inducing a state of 'super-hydrophilicity'. This is demonstrated by the formation of a very thin, uniform film of water on the exposed photocatalytic surface. The water layer then hinders adhesion of external substances to the surface and helps to keep the surface clean.

The combination of a strong destructive oxidation and super-hydrophilicity generated by photocatalytic TiO_2 provides very active surfaces, which not only generate the breakdown of many important gaseous, liquid and particulate pollutants but also significantly reduce the adhesion of all particles (dust) to the surface of the *e*GRC. A significant level of 'self-cleaning' capability is established.

It is possible to identify three mechanisms involved in the self-cleaning process:

1. Particles of organic dirt are broken down in the photocatalytic process.
2. The super-hydrophilicity generated by the activation of the TiO_2 reduces the possibility of adhesion of the unbroken organic pollutants to the surface.
3. Processes (1) and (2) also reduce the opportunity for inorganic particles (dust), unaffected by the photocatalysis, to settle on the surface (especially when vertical), because the surface then contains fewer 'sticky' organic particles to which particles of inorganic dust can adhere.

6.9 Environmental performance

6.9.1 Energy requirements

It is possible to determine the energy requirements for production of the individual raw materials making up GRC. A complete assessment of the energy requirement for production of GRC therefore requires the energy input during its manufacture.

A modern auto-spray traverse system can produce approximately 50 m² / hour of GRC about 12 mm thick. The energy used is approximately 1.7 kWh / 60 kg batch mix, or approximately 29 kWh / tonne of GRC [64].

6.9.2 Environmental impact analysis

A typical product, manufactured both in GRC and in traditional reinforced concrete in large numbers and for an identical purpose, had to be selected so that a reliable, direct comparison and quantification of the difference in environmental impact of both materials could be made [65]. Drainage channels and ducts for communication and other services along railways, as supplied to a typical construction site, were selected. Both products were available in traditional reinforced precast concrete and in a premix type of GRC and are already used in very large numbers.

Environmental impact is assessed by eco-indicators. An example of their application is shown in Tables 6.5 and 6.6. Eco-indicators are determined for the premix GRC drainage channels (3% fibre content) and for cable ducts (3.8% fibre content). The results for the GRC products are taken as the baseline and compared with similar products made of ordinary precast concrete.

The precast concrete drainage channel had a significantly higher impact than the unit made of GRC with 3% of glass fibres. On average, the unweighted impact of the traditional concrete unit was 57% higher, and the weighted one was 75% higher.

Table 6.5 Environmental impact of GRC and precast concrete drainage channels [65]

Environmental impact class	Units	GRC 3% fibre	Precast concrete
Greenhouse gas equivalent	kg CO_2	45.5 (100)	71.7 (158)
Ozone depletion equivalent	mg CFC11	2.00 (100)	6.19 (310)
Acidification equivalent	g SO_4	260 (100)	438 (168)
Eutrophication equivalent	g PO_4	22.2 (100)	41.9 (189)
Heavy metal equivalent	mg Pb	68 (100)	81 (119)
Winter smog equivalent	g	123 (100)	191 (155)
Summer smog equivalent	g C_2H_4	18.9 (100)	30.6 (162)
Primary energy use	MJ	393 (100)	541 (138)
Unweighted average impact	%	100	175
Eco-95 weighted average impact	%	100	157

Note: figures in brackets are percentages of values for GFC.

Table 6.6 Environmental impact of GRC and precast concrete cable ducts [65]

Environmental impact class	Units	GRC 3.8% fibre	Precast concrete
Greenhouse gas equivalent	kg CO_2	13.8 (100)	18.4 (133)
Ozone depletion equivalent	mg CFC11	0.32 (100)	0.99 (309)
Acidification equivalent	g SO_4	66 (100)	109 (165)
Eutrophication equivalent	g PO_4	5.9 (100)	9.8 (166)
Heavy metal equivalent	mg Pb	14.9 (100)	21.3 (143)
Winter smog equivalent	g	34.4 (100)	59.0 (171)
Summer smog equivalent	g C_2H_4	1.23 (100)	7.04 (572)
Primary energy use	MJ	110 (100)	137 (125)
Unweighted average impact	%	100	223
Eco-95 weighted average impact	%	100	161

Note: figures in brackets are percentages of values for GFC.

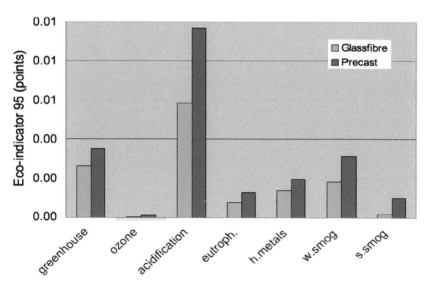

Figure 6.12 Environmental impact parameters: comparison between precast concrete and GRC [65].

The results showed that a precast concrete duct with cover had a significantly higher environmental impact than the same unit made of GRC with 3.8% glass fibres. On average, the unweighted impact of the traditional concrete unit was 123% higher, and the weighted one was 61% higher. Figure 6.12 shows the difference.

The cement content per unit volume of typical GRC is significantly higher than that of normal precast concrete. However, the reduced weight of products with comparable performance made from GRC and the reduced demands on transport also lower the overall environmental impact.

In addition, there are secondary benefits of GRC products due to reduced breakage, reduced amount of site-work and ease of handling. In the specific cases of the two products assessed, GRC shows further non-quantifiable advantages over its precast concrete counterparts. These include:

- Precast concrete channels are more prone to cracking during installation. Replacement units then represent additional environmental impacts in both manufacture and disposal.
- Mechanical handling equipment is not required.
- Installation is quicker, mechanical handling equipment is not required, and the delivered cost is competitive.

In the case of in-ground, track-side ducts, a smaller amount of material is excavated, installation damage is reduced, and track occupancy times are reduced.

6.9.3 Active de-pollution of environment by eGRC

De-pollution of air is caused by the photocatalytic activity of the *e*GRC surfaces [28, 29]. The photocatalytic action is generated by ultraviolet rays (the UV-A part of the spectrum of natural daylight), which interact with the nano-crystalline lattice structure of the anatase version of the titanium dioxide present in the active surface layer of the GRC. In the presence of moisture (water), very highly reactive radicals such as O^{2-} and OH^- are produced on the irradiated surfaces. A broad range of organic and inorganic chemical compounds, both as solids and in the form of liquids or gases, then undergo oxidative destruction when in contact with such strongly photocatalytic surfaces in a complex process.

Trials confirmed [27] that the photocatalytic de-pollution of concrete surfaces in a city-centre busy street was significant, and all showed substantial de-polluting effects. Typical pollutants affected include nitrogen oxides (NO_x), sulphur oxides (SO_x), usually in gaseous form, and numerous organic compounds such as volatile organic compounds (VOCs), formaldehyde, toluene and others in both gas and liquid forms. These are the most frequent and significant constituents of airborne pollution and are known to have adverse effects on human health in densely populated urban centres and industrial zones.

Airborne pollution also directly affects other living organisms such as animals and plants, and its secondary effects include an acceleration of the deterioration of construction materials [66]. The de-pollution has usually been expressed as the observed reduction in concentration of one or more of the pollutants (mostly of the NO_x type) when the quality of the air around the photocatalytic surface was compared with similar control conditions (no photocatalytic concrete surface).

Comparisons between different trials and laboratory test results are very difficult; there are many, very variable influencing factors, such as weather conditions, traffic density and the resultant concentration of specific pollutants, and positioning of the detectors. However, it has been calculated [27] that applying results from practical trials and covering 15% of visible urban surfaces with products with photocatalytic external surfaces in a large city can lead to an overall reduction of the pollution levels by 50%. The de-polluting effects of photocatalytic surfaces are significant (Figure 6.13).

Extensive full-scale trials indicated very substantial reductions of NO_x, which varied according to orientation and the amount of sunshine. Approximately 7000 m² of an asphalt-surfaced, highly trafficked street in Segrate (Italy) was treated with photocatalytic coating (5% TiO_2).

The human toll due to poor air quality is worse than that of road traffic accidents, making it the number one environmental cause of premature death in the EU. It also impacts the quality of life, such as in the case of asthma or respiratory problems. The European Commission is responding with recently adopted new measures to reduce air pollution [66]. The clean air policy package updates existing legislation and further reduces harmful emissions from industry, traffic, energy plants and agriculture, with a view to reducing their impact on human health and the environment. Air pollution also causes lost working days and high healthcare costs, with vulnerable groups such as children,

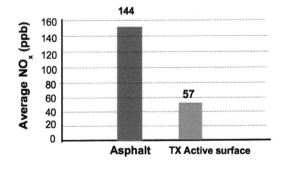

Comparison of NO_x levels

Figure 6.13 De-pollution of air by photocatalysis on a road surface [27].

asthmatics and the elderly being the worst affected. It also damages ecosystems through excess nitrogen pollution (eutrophication) and acid rain [66].

6.9.4 Recycling of GRC

GRC elements are suitable for recycling in a similar way to concrete once any metallic parts of fixings are removed.

Due to its high content of cement, crushed, hardened GRC may still show residual capacity for hydration and contribute to the development of strength of concrete from such a recycled aggregate.

Structural Design | 7

7.1 Principles

The design of GRC elements, which aims to achieve the required performance with maximum efficiency in service and economy in production, requires an integrated approach. This needs to consider

- Optimum shape and size of GRC elements for maximum functionality and an efficient production process within acceptable dimensional tolerances;
- Production process within acceptable dimensional tolerances;
- Ease of de-moulding, handling, storage/stacking and transport, elements without excessively vulnerable, weak parts prone to damage before being erected;
- Fixing points that are durable, minimise internal stressing and are easy to install;
- Handling points facilitating safe transport and installation;
- Service loads and environmental exposure in all probable combinations.

In line with international practice, current structural design of load-bearing GRC elements and systems follows the limit-state approach. The design considers variations in key parameters and properties of the GRC (e.g. flexural strength) and of the structural elements (size, mass, etc.) together with variations in dead and imposed loads of relevant types. The most unfavourable combinations of parameters/characteristics with the same probability are evaluated in terms of both the limit state of failure/collapse and the limit state of serviceability.

Detailed guidance for the design of structural GRC elements, including 'worked examples', is provided by the *Practical Design Guide for GRC using*

Limit State Theory [59] and other publications [55, 56, 66–68]. Large structural design consultancies may have an in-house specialist for GRC design, or a structural engineer with specialist experience of GRC is subcontracted. In smaller projects, structural design can be provided by structural engineers linked with the GRC producer and installer as part of a package.

Joints between GRC panels are typically sealed with an elastomeric sealant such as silicone, urethanes or polysulphides. The chosen sealant should be able to maintain the water-tightness of the joint during movement of the adjoining elements due to dimensional changes and their response to loading. Protection of the sealant from direct sunlight will help to extend its service life. Joints have to be designed to shed and drain water away from the surfaces of the panels.

7.2 Typical structural elements

7.2.1 Single-skin sheet

The simplest element is a panel made of a plain 'single-skin' sheet of GRC, usually 10–15 mm thick. The load-bearing capacity of a simple flat sheet is considerably increased by changing its shape. Figure 7.1 shows typical shapes

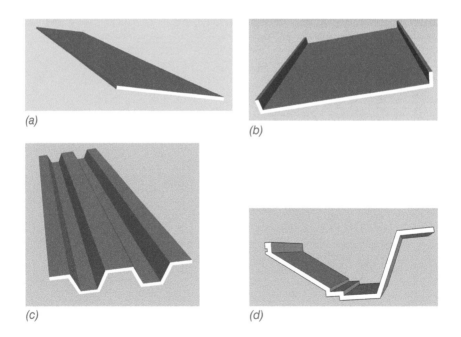

(a)

(b)

(c)

(d)

Figure 7.1 Typical shapes of single-skin GRC panels [59]. (a) Flat sheet; (b) flanged sheet; (c) corrugated sheet; (d) shaped cornice element. (Courtesy of G. Jones.)

formed to give the panel additional stiffness and strength. The most common method to achieve this is by forming edge returns, flanges and ribs.

7.2.2 Ribbed panels

Ribs provide a much greater depth of section and increase load-bearing capacity. Ribs also give the panel a much greater rigidity overall, and perimeter ribs provide in addition good locations for fixings and joints. The typical geometry of a ribbed panel is shown in Figure 7.2. A ribbed panel is produced by spraying additional GRC over rib-formers made of shaped polystyrene or preformed GRC sections.

7.2.3 Sandwich panels

A sandwich panel is created when outer skins of GRC are separated by a core of a light insulating material, all bonded together. Such an arrangement can provide a deep section that resists bending, maximising the contribution of the strength of the GRC itself to the strength of the panel overall. One skin is usually fully in tension while the other one is in compression but these roles can be reversible. Bonding with the core material provides a degree of restraint to a free deflection of the GRC skin and enhances the load-bearing capacity

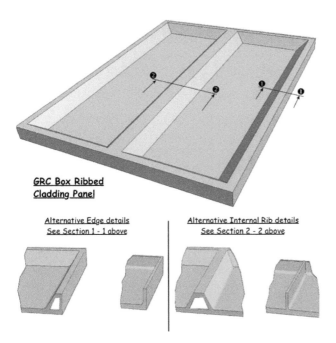

GRC Box Ribbed
Cladding Panel

Alternative Edge details
See Section 1 - 1 above

Alternative Internal Rib details
See Section 2 - 2 above

Figure 7.2 Typical shape of a box-ribbed panel [59].

of the whole unit. In all cases, the core can also provide a very high level of useful thermal insulation.

A *simple form of the sandwich panel* is made of two parallel faces of GRC, usually with edges stiffened by ribs, bonded to the core material (Figure 7.3a). In this case, the core material must have an adequate level of structural strength, as it contributes to the overall performance of the panel by transmitting shear stresses and resisting inter-laminar shear. Being stronger and denser (more rigid) usually means being less effective as thermal insulation compared with 'box-type' sandwich panels. Foamed concrete and rigid polymeric foams can be used as core materials.

The *'box-type' sandwich* (Figure 7.3b) is a more common form. GRC skins are wrapped around the whole of the panel with internal membranes designed to increase the stiffness of the whole element. The space within the 'boxes' is filled with a very lightweight insulating material. Typical materials for the 'core' of such a panel are expanded polystyrene, polyurethane and isocyanate foams, and a very light foamed concrete or a lightweight mix of cement paste with polystyrene beads. All such core materials provide a very high level of thermal insulation.

Due to their geometry/shape, sandwich panels are structurally efficient; however, they are now very rarely used as external cladding of buildings. The reason is that the shape of the cross section of a sandwich panel, particularly of the boxed type, inherently restrains the movement of the GRC skins in response to changes in temperature and/or humidity. High internal stresses

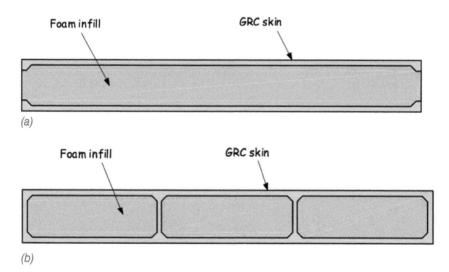

Figure 7.3 Typical shapes of sandwich panels [59]. (a) Sandwich panel, bonded type; (b) sandwich panel, boxed type.

in the GRC are generated, and the panel deforms (e.g. bowing) significantly without any external loads being applied. Moreover, the parallel sides of the panel can each be exposed to very different conditions (e.g. outdoor and indoor environment), and a very steep temperature gradient may develop. This will cause substantial differential movement in respect of each of the sides and result in distortion (e.g. bowing) and cracking of the panel. Panel fixings may not be able to accommodate the effects of such movements, and they tend to contribute to internal over-stress and potential cracking.

7.2.4 Stud-frame system

Very large (up to approximately 25 m² each) GRC cladding panels are usually produced using a 'stud-frame' system. A stud-frame panel is an integral unit, consisting of a steel frame designed to support an external single skin of GRC by fixings which accommodate thermal and moisture-related movements of the composite. The stud-frame panels are manufactured by GRC producers.

Typically, the prefabricated frame of the panel is designed to be stiff but lightweight. This is achieved using thin-walled metallic elements (angles, hollow sections, etc.) made of a durable, corrosion-resistant material. The most common materials are heavily galvanised or stainless steel sections (Figure 7.4).

The GRC skin, which forms the outer surface of the panel, is attached to the frame by appropriately spaced fixings, usually described as flex or gravity anchors, described in Section 7.3.2.

Flex anchors allow for the movement of the GRC and provide support against lateral loads. Gravity anchors transmit the self-weight of the GRC

Figure 7.4 Thin-walled hollow sections in galvanised steel with anchors, ready for the assembly of a stud-frame GRC panel.

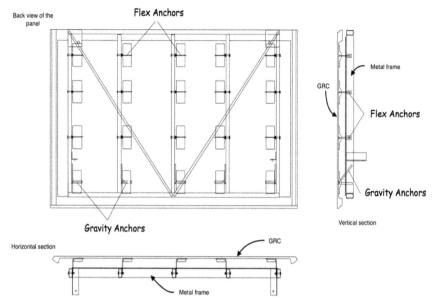

Figure 7.5 Typical arrangement of a GRC stud-frame panel [59].

skin to the stud-frame of the panel. A typical arrangement of a stud-frame GRC panel is shown in Figure 7.5.

The design of the frame in a stud-frame panel also includes fixings for its attachment to the main structure of a building and lifting points for its handling during transport and installation. Stud-frame panels avoid any need to fix GRC itself on a construction site.

The free space between the inner face of the GRC skin and the outside face of a stud-frame element is often filled with a lightweight insulating and fire-resisting material. It is also possible to produce a stud-frame panel with an inner skin, for which GRC or different materials can be used, for example a gypsum plasterboard. Several designs of this type have been developed and successfully used on major projects.

7.3 Fixings

7.3.1 Basic functions and design principles

The fixings of GRC elements must be considered as an integral part of the design process. Guidance can be found in specialist publications [59, 67–69].

Decisions regarding types of fixings have to be taken as early as possible. Fixings carry two main functions for the whole of the lifetime of the structure/building:

- Transmission of imposed loads on GRC elements and of their self-weight to the rest of the structure;
- Simultaneous accommodation of translational and rotational movements caused by dimensional changes (humidity, temperature) and the response of the GRC element to the expected type of load.

An excessive number of fixings (over-fixing) should be avoided as a matter of principle. Fixings that inhibit/restrain the movements of GRC elements generate stresses that lead to cracking and potential failures, as in the case of GRC cladding. The standard approach is to have only four fixing points, preferably in the corners of a vertical panel. A small cladding element would have fixings at the base to carry its weight. Fixings at the top will secure the panel in position and transfer horizontal loads. Multiple fixings are required on large panels, where the size and shape of the element and the supporting building structure dictate their locations.

The structural behaviour of GRC panels, namely deflections, has to be taken into account, as such deformations can generate unacceptable torsional or flexural stresses in the fixings and in the GRC around them. This becomes particularly important when long panels with a large horizontal span are proposed and the ratio of span/depth >4.

The location of fixings governs the distribution of internal stresses in a GRC panel and is of fundamental importance. Such stresses should be minimised by ensuring that loads transferred by fixings are spread over as large an area of the panel as possible. It follows that the thickness of the panel in the vicinity of the fixing may have to be increased, and any direct bearing on GRC needs to be examined.

The location of lifting points in relation to fixings has to be considered in order not to damage fixings during transport and installation. Fixings are not designed, and must not be used, for lifting the panels.

Adequate tolerances must be incorporated into the fixings in order to deal with acceptable inaccuracies of the panels and the supporting structures (as per relevant standards). Ideally, all fixings should be accessible for adjustments to be made, although this is rarely achievable.

The design of all the types of fixings follows the limit state approach using appropriate partial factors.

Precise structural analysis of fixings, particularly of the more complex type, is beyond normal structural design facilities. It means that the performance of a fixing in a full-scale trial panel has to be carried out to verify its ultimate load-bearing capability and behaviour under service loads.

7.3.2 Types of fixings

The most common types are

- Cast-in-socket
- Bonded fixing
- Face fixing

Cast-in-sockets (Figure 7.6a) must be embedded fully in an adequate volume of GRC. The performance of this type of fixing is affected by the edge-distances, the depth of the socket and the type of anchorage (cross-pin, nut in a tapered end) and depends on the type of load transmitted. The top of the socket should be always a few millimetres above the surface of the panel. This will lessen the possible damage due to over-tightening of the fixing against the face of the panel. There are several types of sockets, which differ in how they are anchored. An alternative to cast-in-place sockets is *encapsulated fixings*. Such fixings are very versatile regarding attachments but also require greater amounts of GRC (Figure 7.6b) and have to be made by hand.

Cast-in and encapsulated fixings inherently lead to a substantial thickening of the GRC panel around the fixing point. This has negative effects on both visual performance (potential for 'ghosting') and structural performance (an in-built constraint for shrinkage and other movement).

The introduction of *'wide-headed' fasteners* has been proposed as a way to avoid the problems listed above. In the case of a 'wide-headed' fastener, the embedment and the stress-transfer between the fastener and the GRC occur within the 'normal' thickness of the sheet. A similar approach appears to have been successfully used for fastening thin-walled elements made of other composites, specifically glass fibre reinforced plastic (GRP). There is potential for its use in GRC; however, a considerable amount of research and development will be required to prove its feasibility, assess performance and establish design criteria.

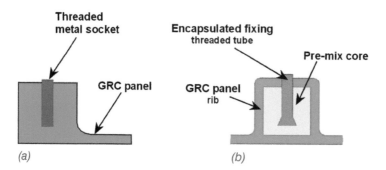

Figure 7.6 Fixing through embedded inserts. (a) Cast-in-socket arrangement; (b) encapsulated fixing. (Courtesy of G. Jones.)

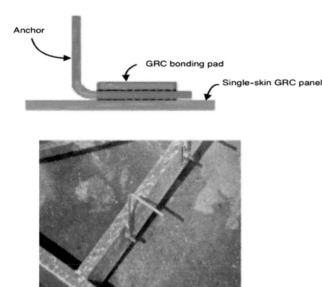

Figure 7.7 Basic arrangement of a bonded fixing. The bonding pad envelops the anchor. (Courtesy of G. Jones.)

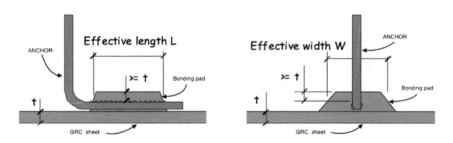

W × L >= 16000 (mm units) & W > 12t

Figure 7.8 Guidance on the minimum size of a bonding pad with a flex anchor [59].

Bonded fixing (Figure 7.7) relies on safe anchorage of the fixing in an adequately large bonding pad provided by additional thickness of the composite over and around the end of the fixing. Bonding pads and fixings are usually attached manually, and care is taken to ensure that the additional fresh GRC interlocks and bonds well with the still fresh body of the panel.

It is essential that the GRC pad and the underlying sheet are well bonded together and that the pad is of an adequate size. Excessive bonding pads waste material and increase the weight of a panel. Guidance on the size of the bonding pad is given in Figure 7.8 [59].

Typical cases of bonded fixings are flex anchors, which transmit horizontal/lateral forces only (Figure 7.9), and gravity anchors, which transmit both horizontal/lateral forces and vertical loads (weight) (Figure 7.10). Both are used to attach a GRC panel to an integral supporting metal frame in the stud-frame system. The stud-frame unit is, in turn, attached to the main structure.

Face fixing is used when there is no practical access to the panel from behind. Such a fixing goes through the panel, and once installed, the external surface is made up from GRC. Examples of a straight face fixing, which includes a cast-in dowel bar, and a 'hidden' face fixing, where the hole is covered by a high-fitting

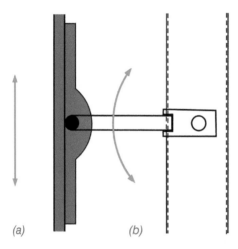

(a)   (b)

Figure 7.9 (a) A typical flex anchor; (b) its degrees of freedom of movement [59].

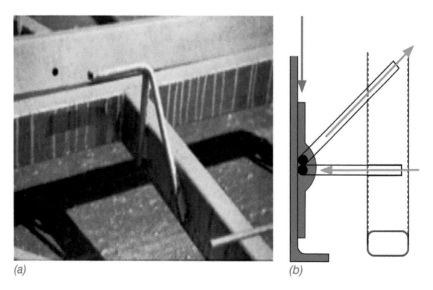

(a)   (b)

Figure 7.10 (a) Gravity anchor attached to the frame; (b) indication of the load paths [59].

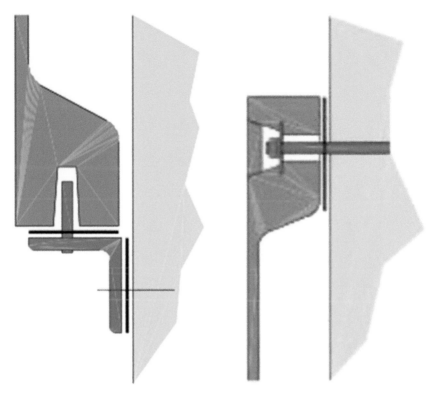

Figure 7.11 Cast-in dowel bar connection. (Courtesy of G. Jones.)

Figure 7.12 'Hidden' face fixing. (Courtest of G. Jones.)

piece of GRC, are shown in Figures 7.11 and 7.12. Even with a precise fit, the location of the fixing will be detectable, unless the surface is painted. If possible, face fixings should be avoided.

Another type of fixing is used when external components have to be attached to a GRC façade element. Such *secondary fixings* are provided by encapsulation of a suitable hardwood strip into the GRC. This allows the external component to be simply screwed in. All such external components must be attached in such a manner that they will not inadvertently restrict the movements that the basic fixings provide for.

7.3.3 Durability

Fixings have to be designed for the lifetime of the structure/building, as it is normally impossible or extremely difficult to inspect or replace them. Whole-life costing should be applied when the economy of the fixings selected is considered.

Stainless steel is the preferred choice of material because of its corrosion resistance; however, it is expensive. The durability of galvanised steel can be adequate, provided the coating can be guaranteed as integral and the required minimum thickness of the zinc coating is maintained over all the surfaces. Other materials used include compound bronzes (phosphorus, aluminium and silicon–aluminium). The choice of material also reflects the service environment/exposure of the fixing (humidity, salt content, etc.).

The material used must always be compatible with that of the supporting frame/structure, avoiding the potential accelerated electrochemical decay, including all metallic parts such as washers, plates, bolts and nuts.

Spacers and packers are made either from stainless steel or from durable polymers. Neoprene washers and packers are used when a degree of movement is to be allowed.

Fixings have to keep joints between panels in the designed position, within acceptable tolerances, to allow appropriate sealants to be applied.

Specification and Compliance | 8

8.1 General guidance, grades and their selection

It is the responsibility of the producer to select a suitable mix design for the product. The mix design must be such that the mechanical properties required for the selected grade of GRC can be consistently achieved.

Target properties for each grade are shown in Tables 6.1 and 6.4. The grades selected must meet the requirements established in the engineering design of the specific product [32].

Two basic categories are recognised: GRC produced by wet curing without the addition of a polymer (e.g. Grade 18) and by dry curing with a polymer added (e.g. Grade 18P).

Grades 8 and 10 are used for premix/spray-premix production of 'non-structural' elements that have to support principally their self-weight alone.

Grade 18 is used for production by traditional spraying. Typical products are cladding panels, where the GRC has to carry wind and other imposed loads in addition to self-weight.

8.2 Production quality control and compliance

Production control of GRC is based on a set of 'deemed-to-satisfy' tests. The values of required parameters and acceptable tolerances are set, and results from production/quality control testing are obtained and compared in a pass/fail manner.

The manufacturer should demonstrate that a quality assurance system is operated. This may be the iGRCA Approved Manufacturer Scheme (AMS), ISO 9001 or similar.

Tolerances for an acceptable variation in properties of finished GRC products, depending on the production process, are given in Table 8.1.

8.3 Sampling and frequency of tests

Tests may be carried out on coupons, ideally cut from the GRC elements to be assessed; however, this is often not feasible. The standard practice is, therefore, to produce a *test board* from which the coupons (test specimens) are cut out. This should be manufactured, de-moulded and cured in the same manner as the GRC element it represents. The quality of the GRC in the test board should be, as far as possible, the same as in the final product. GRC in thicker parts of the product assessed may not be replicated by the test board, the thickness of which is normally limited to 12 mm due to the constraints of typical testing equipment. Test boards must be large enough for a sufficient number of coupons to be cut to meet standard testing requirements. A minimum size of 500 × 800 mm is therefore recommended in the case of a sprayed GRC in order that any directional effects can be identified.

The frequency of production of the test boards shall be not less than one board per day per mixer/pump. This applies to both spray and premix processes. Boards not tested shall be kept for future testing if required. Depending on the process used, the *minimum testing frequency* is

Spray: Twice per week per spray station or for every 10 tonnes of GRC produced

Premix: Once per week per mixer or for every 10 tonnes of GRC produced

These frequencies are the absolute minimum [32, 54]. Individual manufacturers may elect to test more frequently as appropriate for the type and application of the product.

8.4 Testing

Compliance with the properties listed in Tables 8.1 through 8.3 is measured using the tests outlined in this section.

8.4.1 Content of glass fibres

The content of fibres in freshly produced GRC is determined in accordance with either the GRCA *Methods of Testing Glassfibre Reinforced Concrete (GRC) Material* Part 1 [54] or EN 1170 Part 2 or other approved national standards. This test should be carried out on all Grade 18/18P sample boards at least once per day and for Grades 8/8P and 10/10P once per week as a minimum. A description of the test is provided in Appendix B. The content of fibres is not normally measured directly. It is considered adequate provided the tests for strength are acceptable.

Table 8.1 Performance of GRC using spray and premix processes (average values) [32]

	Production process		Test method (EN or GRCA)
	Spray	Premix	
Dry density (kg/m³)	1900 + 300 −200	1900 + 300 −200	EN 1170-6
Strength in bending at 28 days			
LOP (MPa)	8 ± 2	7 ± 2	EN 1170-5
MOR (MPa)	20 ± 5	9 ± 3	
Strain (ε) at MOR (%)	0.8 ± 0.2	≥0.1	
MOR after accelerated ageing (50 immersion/drying cycles) (MPa)	16 ± 4	8 ± 2	EN 1170-8
Ultimate strain (ε) at MOR (%)	≥0.1	≥0.05	EN 1170-5
Water absorption at 24 hours (%)	11 ± 3	11 ± 3	EN 1170-6
Shrinkage/swelling strain (mm/m)	1.2 ± 0.3	1.2 ± 0.3	EN 1170-7
Modulus of elasticity at 28 days (GPa)	10–20	15–25	–

Note: Direct, uniaxial tensile strength is assumed to be typically 50% of LOP in the absence of any other information.

Table 8.2 LOP and MOR of three basic grades of GRC

	Grade		
	8 or 18P	10 or 10P	18 or 18P
Characteristic LOP (MPa)	5	6	7
Characteristic MOR (MPa)	8	10	18

8.4.2 Grade

The grade of the GRC tested is determined from values of LOP and MOR obtained at 7 and/or 28 days in accordance with either the GRCA *Methods of Testing Glassfibre Reinforced Concrete (GRC) Material* Part 3 [54] or EN 1170 Part 5 [58] or with any other approved national standard.

Results from 7 day tests alone are only acceptable if they already exceed the design requirements. Additional information, such as % strain to LOP,

Table 8.3 Values of LOP and MOR for assessment of compliance with specification [32]

	Grade		
	8 or 8P	10 or 10P	18 or 18P
LOP (MPa)			
Mean of four consecutive test board means	7.25	8.00	8.00
Minimum for individual test board mean	5.75	6.00	6.00
MOR (MPa)			
Mean of four consecutive test board means	9.50	12.00	21.00
Minimum for individual test board mean	7.50	8.50	15.00

% strain to MOR and modulus of elasticity, which are often provided automatically by modern test equipment, should be recorded for information only.

As is the established practice with other materials, all the values are 'characteristic', based on a 95% probability of being matched or exceeded in the GRC examined. It is recommended that a minimum of 40 test board results are analysed in the calculation of the 'characteristic' value.

Compliance with the required grade is checked continually by measurement of the key properties (LOP, MOR) of four test specimens in a consecutive order. Mean (average) values are recorded, and compliance is achieved when both the specified averages are always met or exceeded and none of the results falls below the specified minima. The required values are shown in Table 8.3.

It is important to look at the stress-strain or load-displacement curve plotted during a bending test and check whether the shape (e.g. as in Figure 6.5, Chapter 6) corresponds to that obtained in a normal fracture process. The testing equipment may automatically produce LOP and MOR values even when the test procedure and set-up were defective and an abnormal fracture occurred. Using such data would lead to false performance characteristics and a misleading quality assessment.

8.4.3 Bulk density, water absorption and apparent porosity

These are determined in accordance with either the GRCA *Methods of Testing Glassfibre Reinforced Concrete (GRC) Material* Part 2 [54] or BS EN 1170 Part 6 [53] or other approved national standards. It is recommended that such tests are carried out at least once per month.

8.4.4 Other tests

Other tests may be carried out when required by the purchaser, including full-scale load tests of products and components, fire tests, performance tests on cast-in fixings and so on. These tests should always be supervised by the project engineer.

8.5 Dimensional tolerances

Typical GRC products tend to be based on thin-walled elements (10–20 mm), for which a greater degree of accuracy than in precast concrete is required. Regular checking of the thickness, preferably using a purpose-built thickness gauge, is an essential part of the supervision of the production process.

Correct thickness is essential because

- If it is outside the specified + or – tolerances, the product may be rejected.

- Lower thickness exponentially reduces the structural load-bearing capacity of a GRC element.

- Excessive thickness wastes material and increases the weight of the element, which can make it difficult to handle and fit.

- Greater than nominal thickness (an 'over-spray') may be necessary in places such as the corners of ribs and at anchorage points on GRC elements such as cladding panels.

There must be no under-thickness. The thickness of a typical 12 mm GRC sheet element should be within –0 to +6 mm tolerance.

The dimensional tolerances of finished products such as cladding units must comply with the relevant standard for the type of building element (e.g. [70]).

8.6 Potential defects

8.6.1 Uneven fibre content and distribution

Complete uniformity of a 3-D random distribution of the fibre reinforcement in GRC can be achieved only theoretically. It is impossible to achieve in practice, which means that the GRC is always to some extent anisotropic. What matters most, therefore, is the *degree of anisotropy* and the actual prevailing orientation of the fibres in the finished product.

The principal factors that govern the orientation of fibres and the uniformity of their distribution are the production method, the length of fibres and the shape/size/complexity of the mould, which have been discussed in Chapters 5 and 6.

A systematic lack of uniformity of the fibre distribution and/or an inadequate amount of fibres is revealed indirectly by the results of standard flexural tests on specimens cut from a test panel. *Flexural strength lower than expected can be caused either by an inadequate amount of reinforcement or by an excessive amount of fibres.* The latter occurs when the composite is lacking in binder. It is possible for the test panels (simple, flat) to perform well, while the product itself, with a more complex shape, can be defective.

A seriously defective content and distribution of fibres can be also identified by an expert visual inspection. Care must be taken to ensure that the required fibre content is present in all parts of a GRC. Correct fibre content is essential even in sharp returns or in ribs, which are usually impracticable to test but where such a deficiency is more likely to occur with serious consequences. The performance of prototype panels or other products tested for design loads often provides a very useful indication of weak spots and of locations to focus on in production quality control. Visual inspection of completed elements is also useful.

The exact content of glass fibres in hardened GRC can be established by exploiting the ZrO_2 content in the fibres, which separates them from other chemically very similar constituents, namely the aggregate. The analysis requires highly sophisticated equipment available only in specialist, usually research-based laboratories. It is very rarely used.

8.6.2 Inadequate or excessive thickness

The tolerances outlined in Section 8.5 have to be adhered to in order that GRC products are structurally safe and economically produced. Ribs and deep profiles in GRC panels are particularly sensitive to the production process. Examples of defective ribs are shown in sections through GRC panels in Figures 8.1 and 8.2.

Excessive thickness, 'over-thickness', leads to the GRC product being heavier than specified, with several additional potential negative consequences. It also wastes material and increases costs.

8.6.3 Excessive cracking and crazing

A much higher content of the binder in GRC compared with an ordinary mass concrete leads to some crazing and fine cracking of the surface, often affecting just the surface of decorative facing mixes that have no structural function. Such cracking can be caused by a differential in shrinkage strain across the

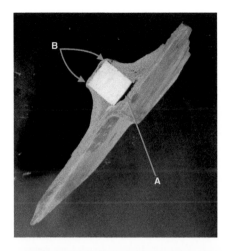

Figure 8.1 A defective rib in a GRC panel. The rib-former (polystyrene) was not adequately held down, and a void 'A' developed. The thickness of the GRC at and between the corners 'B' was also inadequate.

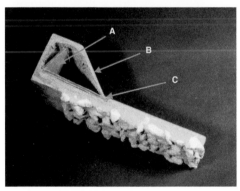

Figure 8.2 Incorrect formation of a rib (A), very inadequate thickness (B) and a stress-concentrating detail (C). Extremely poor-quality workmanship.

thickness of the GRC element. This is normally kept to an acceptable extent in GRC that has been adequately cured (see Section 4.3). Wider cracks indicate excessive deformations, which may be caused by an inadequate structural design (shape of the element, fixings, etc.) or by poor quality GRC.

8.6.4 Porosity, insufficient compaction and surface defects

There is a range of acceptable values for an *apparent porosity*, which reflects internal voids that can be filled by water. It is useful for practical applications. An overall porosity, which includes pores at microscopic scale, is not normally measured, as the tests are sophisticated and testing equipment is expensive. There is very little information available about such porosity and how any values measured could be reliably interpreted and related to the performance of GRC.

An excessive amount of trapped air, showing as an excessive apparent porosity or a lack of compaction (or both), is usually indicated indirectly by a reduced flexural strength of the test specimens.

Care has to be taken to compact the fresh mix and avoid the formation of significant 'blowholes' on the mould side of the product when casting against smooth surfaces. Macroscopic defects of this type are generated by air entrapped at the surface of a mould. Blowholes less than 3 mm in diameter and less than 2 mm deep may be acceptable if their total area is very small; say, less than 1% of the surface. Blowholes should not be discernible if viewed from a distance of more than 5 m from the surface in normal daylight conditions. The choice of an incompatible mould release agent may contribute to the development of such defects.

8.6.5 Excessive efflorescence

Calcium hydroxide and soluble calcium salts generated during the hydration of ordinary GRC matrices can migrate to the external surfaces of GRC elements exposed to cyclic wetting and drying. Carbon dioxide in the air reacts with the salts, and the resulting calcium carbonate may show on the surface as a thin crystalline film, an efflorescence (lime bloom).

Efflorescence is usually seen as an unsightly whitish staining, which is particularly prominent on dark GRC surfaces. Over a period of time, efflorescence itself tends to dissolve, and it is gradually washed away. In GRC applications where external appearance is very important, measures can be taken to reduce the amount of potential efflorescence. Such measures involve lowering the water content of the fresh mix by using admixtures and ensuring good curing conditions. Hydrophobic and pore-blocking surface treatments/sealants may also be specified and may help.

It is also possible to remove the efflorescence by a wash using a weak, usually hydrochloric, acid. This should be tried first in an inconspicuous location, and it needs to be understood that untreated surfaces may then behave and look differently from treated surfaces.

8.6.6 Ghosting

Water is absorbed and evaporates from exposed GRC surfaces, evaporation from thicker sections taking longer. Moist and dry GRC surfaces may show a difference in appearance, with the thicker areas looking darker. This may provide an unsightly, although temporary, appearance of patterns reflecting backing ribs or fixing pads. Ghosting is reduced by making any changes in thickness of external GRC panel skins more gradual.

8.7   Repair and remedial actions

Repairs can be carried out both at the production plant and on site. Any damage to a GRC product has to be assessed first with regard to any potential impairment of its structural performance. If there is no structural damage, patch repairs can be carried out using mixes compatible with the original GRC. Adequate curing of the repair is essential in order that the new material does not shrink away and detach itself from the substrate.

A well-executed repair should not be discernible when observed from a distance greater than 5 metres.

Superficial damage, such as localised chipping, can be repaired on site. It is essential that an adequate curing of the repaired patch is carried out. In general, the slower the hardening of the new material and the lower its shrinkage, the better will be the performance of the repaired element.

Before any repaired GRC element is allowed to be used, all repairs require to be inspected after they have been cured and surface dried.

Health and Safety | 9

Health and safety covers the exposure of workers both to raw constituent materials and to fresh/hardened GRC itself. *Cement* is the main constituent of GRC, and in the UK, the health and safety rules regarding cement in construction [71] have to be observed. Cement can cause health problems associated with skin contact, inhalation and manual handling. Wet GRC can therefore cause dermatitis and burns if in prolonged contact with human skin. Chemical burns can be caused by contact with eyes. Inhalation of cement dust at high levels causes irritation of the nose and throat. Control measures must therefore be employed in the workplace, minimising contact with the skin either directly or indirectly from surfaces contaminated by cement.

Dermatitis caused by cement is controlled by washing the skin with warm water and soap, or another skin cleanser, after exposure and drying the skin afterwards. Gloves may help to protect skin from cement, but they may not be suitable for all types of production work. Caution is advised when using gloves, as cement trapped against the skin inside the glove can cause a cement burn. *Protective clothing, including overalls with long sleeves and long trousers, is advised.* Health surveillance is required for workers who will be working with wet cement-based materials such as fresh GRC on a regular basis to identify as early as possible any indicators of skin changes related to exposure and give early warning of lapses in control. Health surveillance must never be regarded as reducing the need to control exposure or to wash cement off the skin. Simple health surveillance will usually be sufficient. Skin inspections should be done at regular intervals by a competent/responsible person, and the results recorded. A responsible person is someone appointed by the employer who is competent to recognise the signs and symptoms of cement-related dermatitis. The responsible person should report any findings to the employer and will need to refer cases to a suitably qualified person (e.g. an occupational health nurse).

The employer must keep health records containing the particulars, such as those set out in the appendix to the UK General Control of Substances Hazardous to Health (COSHH) Approved Code of Practice [72]. Employees should be encouraged to examine their own skin for any such signs and report them. Reports should be made to the 'responsible person' or to the occupational health nurse. The UK COSHH Regulations require the employer to assess health risks and prevent or control exposure.

Glass fibres as such are not a significant health hazard when handled in the usual manner, unlike some other fibres (specifically asbestos). The size of the fibres (dia. 10–20 μm) used in GRC exceeds the size (3 μm) below which fibres become a health hazard [73]. Glass fibres also do not split lengthwise when broken (e.g. when cutting off cured GRC). Glass fibre, or the GRC dust, does not pose a danger to health; however, it may cause irritation of the skin and respiratory tract, and reasonable precautions should be taken.

Other materials used in production of GRC, such as admixtures, must be stored and handled strictly in accordance with specific instructions and the relevant COSHH requirements.

The measures for protection of health and safety outlined above should be followed even if the country in which the GRC is produced does not require a similar standard of care.

Summary of Benefits | 10

- *Economy through a combination of low weight and high strength*

  Thin-walled GRC elements are usually produced; they are easy to handle, transport and erect. The self-weight of structures decreases when GRC is used, and demands on foundations are reduced. GRC cladding is suitable even for very high-rise buildings and offers good performance under seismic loading.

- *Freedom of shape*

  GRC is easily mouldable into a wide range of shapes, including intricate grilles, panels with double curvature and 3-D objects. The high freedom of shape permits production of structurally very efficient elements. Easily cast, it can produce items with very fine details and reproduce very complex features and elements of both modern and historic buildings.

- *Durability*

  The basic reinforcement is non-ferrous, and GRC products are not susceptible to corrosion as in traditional reinforced concrete. Low permeability and a very slow rate of carbonation offer protection against corrosion of steel in adjacent reinforced concrete. GRC has an inherently high resistance to extreme exposure conditions (freeze/thaw, fire, etc.).

- *Appearance*

  A very wide range of attractive surface finishes is available, satisfying the highest requirements for an aesthetic appearance of new structures and capable of matching the colour and texture of surfaces of existing buildings. Durable and brightly coloured surfaces with enhanced self-cleaning can be achieved in a variety of textures and shapes.

- *Environment*

   The relatively low weight of GRC products reduces CO_2 emissions associated with their transport. There are no volatile organic compounds or other pollutants emitted from the material itself, either in production or in use. GRC is fully recyclable into concrete and other applications. In addition, GRC with photocatalytic surfaces (*e*GRC) directly and significantly reduces the concentration of pollutants in the surrounding air, leading to a better quality of environment, especially in congested urban centres, and at a minimal additional cost.

Applications | 11

11.1 Introduction

GRC has always been a versatile material, however, its range of practical applications has continued to grow, becoming extremely wide.

At one end of the range, it is used for production of small, simple and unsophisticated everyday items (flowerpots, drainage channels, window sills, etc.).

At the other end it has been adopted by leading international architects for large-scale, high-tech iconic projects, coping with the highest demands regarding structural complexity, size of the elements, freedom of shape and spectacular appearance combined with durability and overall quality.

These applications are illustrated using recent photographs and brief data, which indicate the location, purpose/function and completion year of the project, together with the names of the architect and/or structural engineer (if known) and of the producer/supplier of the GRC. Interesting features of each project are also noted.

The use of GRC gave the architects greater degree of freedom of shape, often combined with high demands on environmental performance. Architects worldwide now have a material which makes it possible to realise designs which would have been considered just flights of fancy in the past. Projects shown are only a fraction of what has been built using GRC since the new millennium and what is currently under construction all over the world.

This chapter illustrates the wide range of applications by examples of completed projects in 10 different categories, illustrated in Sections 11.2 through 11.11.

11.2 Mature structures

Projects from the early years of GRC's development, more than 20 years old, when earlier versions of GRC were used.

11.3   Civic/public buildings

Projects where GRC was mostly used as cladding/façade material on important public buildings such as museums, concert halls, government offices and so on.

11.4   Office and commercial buildings

Projects range from standard shopping centres to prestige office buildings of all sizes and shapes.

11.5   Residential buildings

GRC is used in all types of housing, from low-rise housing, small and large private residences to very large blocks of flats and tall residential towers.

11.6   Religious structures

Churches, mosques, temples and similar buildings frequently require highly ornamental and intricate cladding/façades, sometimes replicating historic patterns. GRC has become the material of choice in this area for architects worldwide.

11.7   Art and recreation

The projects reflect the ease with which the GRC elements can be produced in very complex shapes and large sizes without generating weight problems. A full range of colours and surfaces are being exploited.

11.8   Reconstruction/conservation of historic and contemporary buildings

GRC is increasingly used in conservation of built heritage. It is often the best solution when the decayed original elements cannot be reconstructed or replaced but the original appearance has to be maintained. It is equally well suited for the redevelopment of both ancient and modern buildings.

11.9   Interior decoration and furniture

GRC used indoors provides an excellent alternative to other materials (wood, stone, steel, etc.) with its own range of interesting finishes, adaptability to produce curved shapes and additional benefits such as fire resistance and noise abatement.

11.10   Architectural building components

GRC elements used include both tailor-made items for individual building projects and mass-produced ones, including items such as window sills, window surrounds, cornices, screens, balusters and so on. Both premix and sprayed GRC are used.

11.11   Civil and environmental engineering

GRC is a long-established material for permanent formwork (e.g. in bridge construction), often with a dual function as both formwork and a visual surface-former. The latter is currently being used on a very tall residential structure. Relining of decaying sewers with GRC and the use of GRC in drainage exploit the excellent hydraulic properties of GRC surfaces. Highly effective, durable GRC noise barriers are already well-established products.

Small elements include cable ducting and street furniture (bins, benches, planters, etc.).

11.2   Mature GRC structures (>20 years old)

Photo courtesy of C. Rickard

Location:	30 Canon Street, London, UK
Year of completion:	1974
Purpose/function:	Office building, originally for the Crédit Lyonnais bank in the City of London
Architect/Engineer:	Whinney, Son & Austen Hall/Ove Arup
GRC producer:	Portland Group

Notes: This is one of the very first major applications of an early version of GRC as the cladding of a large building. The façade was made up of approximately 1,900 double-sided hollow GRC panels. The spray-up process was used. A comprehensive structural inspection was carried out in 2002 including tests for flexural strength (EN 1170:5) on specimens removed from the panels. A loss of ductility was observed with MOR reduced by approximately 50%, less than expected for 28-year-old external elements made of an early version of GRC. The MOR was not expected to decrease further with time. Long-term performance of the GRC may have been improved by an application of a water-repellent treatment to external surfaces of the panels. The original design was based on fixings which allowed movement and prevented the panels from being internally stressed. The façade as shown on this photo has been coated and it has recently been cleaned.

Photo courtesy of N. Brand

Location: Ådalsparken, Hoersholm, Denmark
Year of completion: 1991–1993/1994
Purpose/function: Residential and hotel development
Architect/Engineer: Jorsal, Bech & Thomsen A/S
GRC producer: BB fiberbeton A/S, Denmark

Notes: Three very large buildings are the core of the Ådalsparken development.

Photo courtesy of G. T. Gilbert

Location: San Francisco, California, USA
Year of completion: 1984
Purpose/function: Hotel at Parc Fifty Five and residential units
Architect/Engineer: Daniel Mann, Johnson & Mendenhall
GRC producer: Lafayette Manufacturing, Inc. Hayward, USA

Notes: 17,000 m² of GRC was used.

Location:	San Francisco, California, USA
Year of completion:	1989
Purpose/function:	Hotel
Architect/Engineer:	Zeidler Partnership Architects and A J Lumsden DMJM
GRC producer:	Lafayette Manufacturing, Inc. Hayward, USA

Notes: 32,000 m² of GRC was used to produce 2,400 panels. The 133 m tall, 39-storey hotel was completed and opened just before a major earthquake later in the day. The hotel performed well, it survived with just one window being lost.

11.3   Civic/public buildings

Photo courtesy of Cem–Fil/OCR

Location:	Recital Centre, Melbourne, Australia
Year of completion:	2009
Purpose/function:	Concert hall/performing arts centre
Architect/Engineer:	Ashton Raggatt McDougal
GRC producer:	Asurco Contracting Pty Ltd

Photo courtesy of Fibrobeton

Location:	Kutaisi, Georgia
Year of completion:	2012
Purpose/function:	House of Parliament
Architect/Engineer:	Mamoru Kawaguchi and GMD Ingenieros
GRC producer:	Fibrobeton, Turkey

Notes: 12,000 m^2 of GRC using Fibro-Multiform® Technology was used.

Photo courtesy of Nanjing Beilida Co.

Location: Wuhan, China
Year of completion: 2011
Purpose/function: Museum of the 1911 Revolution
GRC producer: Nanjing Beilida Co.

Notes: The first very large, large-scale use of photocatalytic eGRC for the whole façade (approximately 35,000 m^2). Very large panels replicating rough-hewn red granite were used.

Photo courtesy of C. Rickard

Location: Sydney, Australia
Year of completion: 2013
Purpose/function: Museum of Contemporary Art – extension
Architect/Engineer: Charles Rickard
GRC producer: Precast Concrete, Brisbane

Notes: Very large, 9 m × 3 m GRC panels with returns were used.

Photo courtesy of Nanjing Beilida Co.

Location: Nanjing, China
Year of completion: 2014
Purpose/function: Nanjing Youth Olympic Centre – hotel and conference centre
Architect/Engineer: Zaha Hadid and Nanjing Beilida New Materials System Engineering Co. Ltd
GRC producer: Nanjing Beilida Co. Ltd

Notes: A total of 110,000 m² of GRC was used in over 12,000 elements, classified into four types: 40,000 m² of flat panels, normal size of 3 m × 2 m up to 6 m × 3 m; 32,000 m² of folded panels, normal size of (2+2) m × 2 m up to (5+2) m × 3 m; 30,000 m² of double-curved panels, normal size of 3 m × 2 m up to 6 m × 4 m; 8,000 m² of single-curved panels, normal size of 3 m × 2 m up to 6 m × 4 m. The panels were designed to be installed in an inclined line both for roof and walls.

Photo courtesy of C. Rickard

Location: Geelong, Australia
Year of completion: 2015
Purpose/function: Central Library
Architect/Engineer: ARM Architecture
GRC producer: Asurco Contracting Pty Ltd

Notes: An award-winning building. The dome is made up of 300 GRC elements.

Photo courtesy of Nanjing Beilida Co. Ltd

Location: Yinchuan, China
Year of completion: 2015
Purpose/function: Art Gallery
Architect/Engineer: WAA Architecture and Interior Design/Nanjing Beilida New Materials System Engineering Co. Ltd
GRC producer: Nanjing Beilida Co. Ltd

Notes: A total of 12,000 m² of GRC was used, with both single- and double-curved panels. The longest element was 18 m long.

Photo courtesy of iGRCA

Location: Gori, Georgia
Year of completion: 2013
Purpose/function: Government building, Justice House
Architect/Engineer: Irakli Sarasidze
GRC producer: Fibrobeton

Photo courtesy of Nanjing Beilida Co. Ltd

Location: Nanjing, China
Year of completion: 2014
Purpose/function: Museum of Tangshan Pithecanthropus
Architect/Engineer: Nanjing Yangtze Metropolitan Architectural Design Co. Ltd
Nanjing Beilida New Materials System Engineering Co. Ltd
GRC producer: Nanjing Beilida Co. Ltd

Notes: A total of 9,000 m² of GRC was used to create imitation stone texture with different performances. The flowing GRC curtain wall provided a distinct visual impact displaying interesting details.

Photo courtesy of Nanjing Beilida Co. Ltd

Location:	Shanghai, China
Year of completion:	2011
Purpose/function:	Qian Xuesen Museum
Architect/Engineer:	He Jingtang Design Team of South China University of Technology and Nanjing Beilida New Materials System Engineering Co. Ltd
GRC producer:	Nanjing Beilida Co. Ltd

Notes: A total of 6,350 m² of GRC was used. Deeply profiled, rough-textured GRC was used, with a portrait imprinted on the main façade.

Photo courtesy of Nanjing Beilida Co. Ltd

Location: Jiujiang, China
Year of completion: 2016, under construction
Purpose/function: Culture and art centre
Architect/Engineer: Architects & Engineering Co. Ltd of Southeast University
                           Nanjing Beilida New Materials System Engineering Co. Ltd
GRC producer: Nanjing Beilida Co. Ltd

Notes: A total of 17,000 m² of GRC was used. BIM technique and advanced three-dimensional engineering software were used to ensure accurate match and flexible docking of this typical non-linear structure.

11.4 Office and commercial buildings

Photo courtesy of BB fiberbeton

Location:	Copenhagen, Denmark
Year of completion:	2016
Purpose/function:	University offices and research laboratories
Architect/Engineer:	C.F. Møller Architects A/S & Wagner-Biro Stahlbau Ag N. H. Hansen & Søn A/S and Elindco Byggefirma A/S
GRC producer:	BB fiberbeton A/S, Denmark

Notes: A mix of stud-frame panels with embedded steel brackets and recessed holes. 8,000 elements with hundreds of variations, approximately 11,000 m² of GRC in total.

Photo courtesy of Karma Lama

Location: Thimphu, Bhutan
Year of completion: 2015
Purpose/function: Hotel
GRC producer: Bhutan GRC/Peljorkhang Pvt. Ltd

Photo courtesy of Dima Pushesh

Location: Tehran, Iran
Year of completion: 2016
Purpose/function: Heravi Shopping Centre
Architect/Engineer: Taraghi Jah
GRC producer: Dima Pushesh Co. Ltd

Photo courtesy of iGRCA

Location: Istanbul
Year of completion: 2014
Purpose/function: Business centre
Architect/Engineer: Tago Architecture
GRC producer: Fibrobeton, Turkey

Photo courtesy of iGRCA Detail of façade

Location:	New Ludgate, London, UK
Year of completion:	2015
Purpose/function:	Offices and commercial retail space
Architect/Engineer:	Fletcher Priest Architects and Sauerbruch Hutton Architects
Main contractor:	Skanska
GRC producer:	Techcrete

Notes: The City of London Building of the Year 2016. The 'nose-shaped' panels were pre-fixed to unitised windows. GRC surfaces were acid etched. 1,225 panels 1.5 m to 3 m long and 2,244 m of ledges were installed.

Photo courtesy of iGRCA

Location: Bradford, UK
Year of completion: 2015
Purpose/function: University of Bradford, offices and laboratories
Architect/Engineer: Farrell and Clark architects
GRC producer: GB Architectural Products Ltd

Photo courtesy of BB fiberbeton

Location: Lund, Sweden
Year of completion: 2014
Purpose/function: Kristallen, office block
Architect/Engineer: Christensen & Co Arkitekter A/S, REKAB
Entreprenad AB STATICUS UAB
GRC producer: BB fiberbeton A/S, Denmark

Photos courtesy of iGRCA & Fibrobeton

Location: Gaziantep, Turkey
Year of completion: 2013
Purpose/function: Shopping Centre 'Prime Mall'
Architect/Engineer: Erginoğlu & Çalışlar Architects
GRC producer: Fibrobeton, Turkey

Notes: 11,520 m² of heat insulated Fibrofombeton® GRC panels were used.

Photo courtesy of Ian White

Location: Poland
Year of completion: 2015
Purpose/function: Commercial building

Note: The cladding is made of photocatalytic eGRC.

11.5   Residential buildings and developments

Photo courtesy of GB Architectural Cladding Products Ltd

Location: Union Street, Southwark, London, UK
Year of completion: 2014
Purpose/function: Private residential development
Architect/Engineer: Smart Crosby International specialist GRC consult-
                        ants & designers, Mount Anvil Developments Ltd,
                        English Architectural Glazing (EAG)
GRC producer: GB Architectural Cladding Products Ltd

Notes: Approximately 2,500 m² of GRC cladding was used.

Photo courtesy of N. Brandt

Location: Virum, Denmark
Year of completion: 2016 (stage 1 completed, project continues)
Purpose/function: Residential development 'Sorgenfrivang'
Architect/Engineer: Domus Arkitekter A/S / NCC Construction A/S
GRC producer: BB fiberbeton A/S, Denmark

Notes: Approximately 13,000 GRC elements were used in a mixture of stud-frame panels and panels joined by embedded fixings. 15-story façade of 20,000 m² of GRC.

Photo courtesy of Fibrobeton

Location: Istanbul, Turkey
Year of completion: 2011
Purpose/function: Residential development 'Anthill' including a hotel
Architect/Engineer: MM Proje
GRC producer: Fibrobeton, Turkey

Notes: Two tower blocks, 210 m tall, 54 story, 64,000 m² of GRC panels.

Photo courtesy of Nanjing Beilida Co. Ltd

Location: Tangshan City, China
Year of completion: 2014
Purpose/function: The Third Space, residential building
Architect/Engineer: Li Xinggang Office of China Architecture Design Institute/Nanjing Beilida New Materials System Engineering Co. Ltd
GRC producer: Nanjing Beilida Co. Ltd

Notes: A total of 12,000 m² of GRC was used; the building envelope was covered with heat retaining GRC panels.

Photo courtesy of iGRCA

Location: Moscow
Year of completion: 2011
Purpose/function: Private residence
Architect/Engineer: Zaha Hadid Architects
GRC producer: Fibrobeton, Turkey

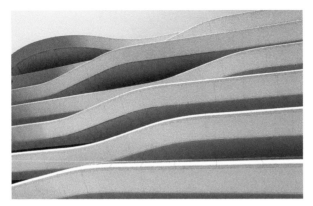

Photos courtesy of Beni Cohen

Location:	Istanbul, Turkey
Year of completion:	2012
Purpose/function:	Residential tower blocks 'Ikon'
Architect/Engineer:	Tago Architects
GRC producer:	Fibrobeton, Turkey

Photo courtesy of Yanfei Che

Location: Shunyi, Beijing, China
Year of completion: 2015
Purpose/function: Private residence
Architect/Engineer: Xiang Chen
GRC producer: Zhongke Masons Building Co.

 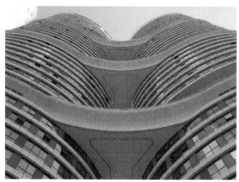

Photos courtesy of Beni Cohen

Location: Istanbul, Turkey
Year of completion: 2016
Purpose/function: Residential development 'Petek'
Architect/Engineer: Tago Architecture
GRC producer: Fibrobeton, Turkey

Photo courtesy of Nanjing Beilida Co. Ltd

Location: Pak Shek Kok, HongKong, China
Year of completion: 2014
Purpose/function: Residential and commercial building
Architect/Engineer: Foster+Partners Architecture Design Office/Nanjing Beilida New Materials System Engineering Co. Ltd
GRC producer: Nanjing Beilida Co. Ltd

Notes: A total of 8,000 m² of GRC was used. Production, assembly and installation of precast concrete and GRC decorative components were completed at the same time in a new construction system for building exterior walls.

Photo courtesy of Beni Cohen

Location:	Istanbul
Year of completion:	2011
Purpose/function:	Multipurpose residential development
Architect/Engineer:	Evrenol Architects
GRC producer:	Fibrobeton, Turkey

11.6 Religious structures

Photo courtesy of Karma Lama

Location:	Takela, Eastern Bhutan
Year of completion:	2015
Purpose/function:	'Golden throne' as a base for a 47 m tall statue of Padmasambhadva
Architect/Engineer:	Civil-Park International, Bangkok, Thailand
GRC producer:	Bhutan GRC/Peljorkhang Pvt. Ltd

Notes: 2300 m² of GRC panels and artwork.

Photo courtesy of iGRCA

Location: Kyiv, Ukraine
Year of completion: 2011
Purpose/function: Cathedral of the Resurrection of Jesus Christ
Architect/Engineer: Nicola Levchuk

Notes: The main domes were made of four large GRC elements, each approximately 80 m², produced in situ inside the building.

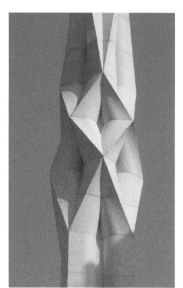

Photo courtesy of iGRCA

Location: Istanbul, Turkey
Year of completion: 2015
Purpose/function: Mosque
Architect/Engineer: Hassa Architecture
GRC producer: Fibrobeton, Turkey

Notes: Photo on the right: detail of the minaret clad in GRC.

Photo courtesy of Karma Lama

Location: Bhutan
Year of completion: 2015
Purpose/function: 108 No of Stupas/Choetens at Kenchongsum Lhakhang monastery
GRC producer: Bhutan GRC/Peljorkhang Pvt. Ltd

11.7 Art and recreation

Photo courtesy of C. Rickard

Location: Sentosa Island Park, Singapore
Year of completion: 1996
Purpose/function: The 'Merlion', Symbol of Singapore
Architect/Engineer: Charles Rickard
GRC producer: Glenn Industries, Caravelle Constructions

Notes: A 37 m tall, 'free-form' structure with a lift inside to an observation platform at the top.

Photos courtesy of Nanjing Beilida Co. Ltd

Location: Changzhou, China
Year of completion: 2015
Purpose/function: Dino Water Town – Leisure Centre
Architect/Engineer: F+S USA / Nanjing Beilida New Materials System
                           Engineering Co. Ltd
GRC producer: Nanjing Beilida Co. Ltd

Notes: A total of 93,000 m² of GRC was used in a large dinosaur theme park in East China. Multicoloured GRC was used as the curtain wall system for the entire façade of the building.

Photo courtesy of C. Rickard

Location: Goulbourn, Australia
Year of completion: 1985
Purpose/function: Symbol of the Merino wool industry
Architect/Engineer: Charles Rickard
GRC producer: Mokany Bros.

Photo courtesy of C. Rickard

Location: Melbourne, Australia
Year of completion: 2015
Purpose/function: Monument Park, Festival of Sculpture
Architect/Engineer: Callum Morton

Location: Bhutan
Purpose/function: Base for the National Pole
GRC producer: Bhutan GRC/Peljorkhang Pvt. Ltd

Photo courtesy of Yanfei Che

Location: Changping, Beijing, China
Architect: Xian Cheng
GRC producer: Zhongke Masons Building Co.

Notes: GRC used to simulate ancient wood.

11.8 Reconstruction/conservation of historic and contemporary buildings

Photo courtesy of iGRCA

Location:	The City College of New York, New York, USA
Year of completion:	2013
Purpose/function:	Renovation of the iconic Shepard Hall building
Architect/Engineer:	Carl Stein, Elemental Architects LLC
GRC producer:	Several manufacturers

Notes: More than 72,000 pieces of GRC, including 4,000 finite sculptures, were used to restore the complex façade by replacing the original glazed terracotta elements. The project took almost 20 years to complete and received numerous awards.

Façade of the building before redevelopment

The same building after redevelopment

Photo courtesy of Ben Burge

Location:	Rothley, Leicestershire, UK
Year of completion:	2007
Purpose/function:	Redevelopment of a residential property (A1 Alpha Properties)
GRC Producer:	Haddonstone, UK

Notes: Approximately 130 t of GRC was used.

Photo courtesy of Hani Jhundi

Location:	Jeddah, Kingdom of Saudi Arabia
Year of completion:	2016
Purpose/function:	Galleria Hotel
Architect/Engineer:	Kling Consult
GRC Producer:	Station Group LLC

Notes: A replica of a late nineteenth-century style of building.

11.9   Interior decoration and furniture

Photo courtesy of iGRCA

Location:	Kingdom of Saudi Arabia
GRC producer:	Arabian Tile Company Ltd

Photo courtesy of C. Rickard

Location: Australia
Purpose/function: Planter and seating
Architect/Engineer: C. Rickard

Photo courtesy of N. Brandt

Location: Uppsala, Sweden
Year of completion: 2007
Purpose/ function: Concert hall, acoustic lining of walls
Architect/Engineer: Henning Larsen Architects
GRC producer: BB fibrobeton A/S, Denmark

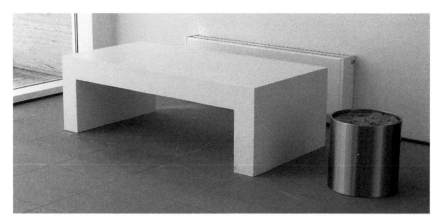

Photo courtesy of N. Brandt

Purpose/function: Bench
GRC producer: BB fibrobeton A/S, Denmark

Photo courtesy of N. Brandt

Purpose/function: Airport counter and seating
GRC producer: BB fibrobeton A/S, Denmark

Photo courtesy of Bhutan GRC

Location: Thimphu, Bhutan
Year of completion: 2015
Purpose/function: Balustrades and columns, interior of a holiday resort
GRC producer: Bhutan GRC/Peljorkhang Pvt. Ltd

Photo courtesy of N. Brandt

Purpose/function: Reception desk
GRC producer: BB fibrobeton A/S, Denmark

Photo courtesy of N. Brandt

Purpose/function: Hand washbasin
GRC producer: BB fibrobeton A/S, Denmark

Photo courtesy of Yanfei Che

Location: Huangpu, Shanghai, China
Purpose/function: Lining of walls
Architect/Engineer: Yiling Ma, East China Architectural Design and
                        Research Institute
GRC producer: Shanghai Wuzhengsheng Decoration Engineering

11.10   Architectural building components

Photo courtesy of C. Rickard

Location: Melbourne, Australia
Year of completion: 2014
Purpose/function: Swinburne University of Technology, Sunscreens
Architect/Engineer: C. Rickard
GRC producer: Minesco

Photo courtesy of Bhutan GRC

Location: Bhutan
Purpose/function: A window surround in traditional Bhutanese style
GRC producer: Bhutan GRC/Peljorkhang Pvt. Ltd

Photo courtesy of iGRCA

Location: Isfahan, Iran
Year of completion: 2014
Purpose/function: Façade elements, Isfahan International Public Gatherings Centre
Architect/Engineer: Atkins
GRC producer: Deesman Company of Tehran Road, Isfahan, Iran

Notes: 8500 m² of GRC was used.

Photo courtesy of Dima Pushesh

Location: Heravi Shopping Centre, Tehran, Iran
Year of completion: 2016
Purpose/function: Façade element
Architect/Engineer: Taraghi Jah
GRC producer: Dima Pushesh Co. Iran

Photo courtesy of Beni Cohen

Location: Ankara, Turkey
Year of completion: 2016
Purpose/function: Façade element/sunscreen
Architect: AW Architects (USA)
GRC producer: Fibrobeton, Turkey

Photo courtesy of G.T. Gilbert

Location: San Francisco, USA
Year of completion: 1984
Purpose/function: Column covers
Architect/Engineer: Daniel Mann, Johnson & Mendenhall
GRC producer: Lafayette Manufacturing, Inc. Hayward, USA

11.11 Civil and environmental engineering

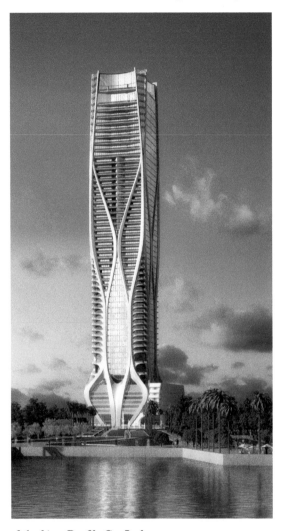

Photo courtesy of Arabian Profile Co. Ltd

Location: 1000 Museum, Miami, Florida, USA
Year of completion: 2015–2018 (under construction)
Purpose/function: Residential development
Architect/Engineer: Zaha Hadid Architects, e-construct
GRC producer: Arabian Profile Co. Ltd

Notes: A 300-m tall, 62-story building. GRC panels serve both as a permanent formwork for reinforced concrete and providing simultaneously high-quality exposed finished surfaces. Approximately 17,000 m² of the dual-purpose GRC.

Photo courtesy of P. Ridd

Purpose/function: Ribbed GRC panels used as *permanent formwork* in concrete bridge construction

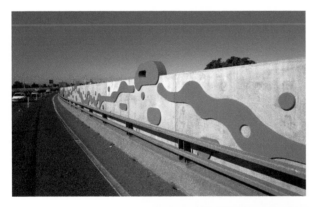

Photos courtesy of P. Ridd
Location: Australia and Hong Kong
Purpose/function: Noise barriers

Photo courtesy of P. Ridd

Location: Malaysia
Purpose/function: Noise barriers

Photo courtesy of P. Leber

Location: Modrice, Czech Republic
Purpose/function: Noise barrier
GRC producer: DAKO Brno s r.o.

Photo courtesy of P. Ridd

Purpose/function: GRC used as lining of irrigation channels in a developing country

Photo courtesy of Althon Ltd

Stacked GRC drainage channel units

Photo courtesy of Althon Ltd

GRC units:	SWALE drainage inlet and the channel with cover, already installed
Location:	UK
Purpose/function:	Drainage channels
GRC producer:	Althon, UK

Photo courtesy of Yanfei Che

Location: Shanxi province, China
Purpose/function: Rubbish bins
Architect: Beijing Central Academy of Art
GRC producer: Zhongke Masons Building Co.

Photo courtesy of C. Rickard

Purpose/function: Drainage boxes
GRC producer: Mascot, Australia

Standards | 12

European standards (EN), which also apply in Great Britain (published by the BSI), and US standards (published by ASTM) relevant to GRC are widely used and are listed here. Where appropriate, these are also mentioned in the text and in the references.

Standards are continually updated; a check for the latest versions should be carried out before use. The list provided below is not exhaustive; it concentrates on the standards most in use internationally.

A number of countries have their own national standards. These do not necessarily replicate the international standards, but it may be necessary to comply with them.

The International Glassfibre Reinforced Concrete Association also publishes documents, which can be used in lieu of or to complement the international/national standards.

1 European Standard EN 1169:1999 (British BS EN)
   Precast concrete products. General rules for factory production control of glassfibre reinforced cement

2 EN 1170 Part 1:1998: Precast concrete products. Test method for glassfibre reinforced cement. Measuring the consistency of the matrix. 'Slump test' method
   Part 1: Measuring the plasticity of the mortar. 'Slump test' method
   Part 2: Measuring the fibre content in fresh GRC. 'Wash-out test'
   Part 3: Measuring the fibre content of sprayed GRC
   Part 4: Measuring bending strength. 'Simplified bending test' method
   Part 5: Measuring bending strength. 'Complete bending test' method
   Part 6: Determination of the absorption of water by immersion and determination of the dry density

Part 7: Measurement of extremes of dimensional variations due to moisture content
Part 8: Cyclic weathering type test

3 European Standard EN 14649:2005 (British BS EN)
Precast concrete products. Test method for strength retention of glass fibres in cement and concrete (SIC test)

4 European Standard EN 15422:2008 (British BS EN)
Precast concrete products. Specification of glass fibres for reinforcement of mortars and concretes

5 European Standard BS EN 15191:2009 (British BS EN)
Precast concrete products. Classification of glassfibre reinforced concrete performance

6 American Society for Testing and Materials: ASTM C974-03
Standard Test Method for Flexural Properties of Thin-Section Glass-Fiber-Reinforced Concrete (Using Simple Beam With Third-Point Loading). ASTM, Pennsylvania, USA, 2003

7 American Society for Testing and Materials: ASTM C948-81
Standard Test Method for Dry and Wet Bulk Density, Water Absorption, and Apparent Porosity of Thin Sections of Glass-Fiber Reinforced Concrete. ASTM, Pennsylvania, USA, 2001

8 American Society for Testing and Materials: ASTM C1228-96
Standard Practice for Preparing Coupons for Flexural and Washout Tests on Glass Fiber Reinforced Concrete. ASTM, Pennsylvania, USA, 2004

9 American Society for Testing and Materials: ASTM C1229-94
Standard Test Method for Determination of Glass Fiber Content in Glass Fiber Reinforced Concrete (GFRC) (Wash-Out Test). ASTM, Pennsylvania, USA, 2001

10 American Society for Testing and Materials: ASTM C1230-96
Standard Test Method for Performing Tension Tests on Glass-Fiber Reinforced Concrete (GFRC) Bonding Pads. ASTM, Pennsylvania, USA, 2004

11 American Society for Testing and Materials: ASTM C1560-03
Standard Test Method for Hot Water Accelerated Aging of Glass-Fiber Reinforced Cement-Based Composites. ASTM, Pennsylvania, USA, 2003

References | 13

1 Biryukovich K.L., Biryukovich Yu.L.: *Steklocement* (in Russian), Budivelnik, Kiev, 1964.

2 Grimer F.J., Ali M.A.: The strength of cements reinforced with glass fibres, *Magazine of Concrete Research*, Vol. 21, No. 16, March 1969, pp. 23–30.

3 Allen H.G.: Fabrication and properties of glass reinforced cement, *Composites*, September 1969, pp. 19–24.

4 Tallentire A.G., Majumdar A.J.: Glassfibre reinforced cement base materials, in *American Concrete Institute Special Publication*, No. 44, January 1974, pp. 351–362.

5 GRCA: Glassfibre Reinforced Cement, *Proceedings of the 1st International Congress*, Brighton, UK, October 1977.

6 Fordyce M.W., Wodehouse R.G.: *GRC and Buildings*, Butterworths, London, 1983.

7 True G.: *GRC: Production and Uses*, Routledge, UK, January 1985, 110 p.

8 Majumdar A.J., Laws V.: *Glassfibre Reinforced Cement*, BSP Professional Books, Oxford, 1991.

9 Bartos P.: *GRC: Investigation of Bond between Reinforcement and Hardened Cement Paste*, Report, Department of Civil Engineering, University of Southampton, March 1970.

10 Brameshuber W., Hegger J., Will N. (Eds.): *Textile Reinforced Concrete – State-of-the-Art*, RILEM Publications s.a.r.l., Cachan, 2006.

11 Curbach M., Jesse F.: High-performance textile-reinforced concrete, *Structural Engineering International*, Vol. 9, No. 4, IABSE, November 1999, pp. 289–291.

12 Bijen J., Jacobs M.J.N.: Properties of glass fiber reinforced, polymer modified cement, *Journal of Materials and Structures*, Vol. 15, No. 89, September–October 1981, pp. 445–452.

13 Ball H.P., Jr.: 25 years of Forton Polymer Modified GFRC: The reasons for its use, *Proceedings of the 14th GRCA Congress*, Hong Kong, 2005, International Glass Reinforced Concrete Association, Camberley, UK.

14 Gomes C.E.M., Savastano H. Jr.: Study of hygral behaviour of non-asbestos fibre cement made by similar Hatschek process, *Materials Research*, Vol. 7, No. 1, 2014, pp. 121–129.

15 Page C.L., Short N.L., Purnell P.: Super-critical carbonation of glass fibre-reinforced cement. Part 1: mechanical testing and chemical analysis, *Composites Part A: Applied Science and Manufacturing*, Vol. 32, No. 12, December 2001, pp. 1777–1788.

16 Majumdar A.J., Singh B.: GRC made from supersulphated cement: 10 year results, *Composites*, Vol. 18, No. 4, September 1987, pp. 329–333.

17 Diao J., Mingfang C., Dongyou Q., Yunbei L., Shujiang Y.: Low-alkalinity sul-phoaluminate cement and its application in GRC products in China, *Proceedings of 13th GRCA Congress*, Dublin, 2001, Clarke N. (Ed.), Camberley, pp. 345–354.

18 Litherland K., Oakley D., Proctor B.: The use of accelerated ageing procedures to predict the long term strength of GRC composites, *Cement and Concrete Research*, Vol. 11, 1981, pp. 455–466.

19 Litherland K.L., Maguire P., Proctor B.A.: A test method for strength of glass fibres in cement, *International Journal of Cement Composites and Lightweight Concrete*, 1984, Vol. 6, pp. 39–45.

20 EN 14649:2005: Precast concrete products. Test method for strength retention of glass fibres in cement and concrete (SIC test).

21 EN 15422:2008: Precast concrete products. Specification of glass fibres for rein-forcement of mortars and concretes.

22 ASTM C1666/C1666M - 08: Standard Specification for Alkalli-Resistant (AR) Glass Fiber for GFRC and Fiber-Reinforced Concrete and Cement, USA, 2015.

23 Scheffler C., Gao S.L., Plonka R., Mader E., Hempel S., Butler M., Mechtcherine V.: Interphase modification of alkali-resistant glass fibres and carbon fibres for textile reinforced concrete, *Composites Science and Technology*, Vol. 69, 2009, pp. 531–524.

24 Van Itterbeeck P., Cuypers H., Orlowsky J., Wastiels J.: Evaluation of the strand in cement (SIC) test for GRCs with improved durability, *Materials and Structures*, Vol. 41, No. 6, July 2008, pp. 1109–1116.

25 Gilbert G.T.: The use of admixtures for GRC production, International GRCA techNOTE, *Concrete*, October 2008, pp. 7–9.

26 De Schutter G., Bartos P.J.M., Gibbs J., Domone P.: *Self-Compacting Concrete*, Whittles Publishing/CRC Press, Dunbeath, UK, 2008.

27 Cassar L.: Nanotechnology and photocatalysis in cementitious materials, in *Nanotechnology in Construction – NICOM2*, *Proceedings of the 2nd International Symposium on Nanotechnology and Construction*, de Miguel Y.R., Porro, A., Bartos P.J.M. (Eds.), RILEM Publications s.a.r.l., 2006.

28 Bartos P.J.M.: *e*-GRC: Improving appearance of concrete buildings and quality of urban environment, *Concrete*, Vol. 43, No. 3, April 2009, pp. 25–28.

29 Bartos P.J.M., Bloomer S., Higgins D.: *Photocatalytic Concrete*, The Concrete Society, Camberley, UK, CPS 157, Pt. 1 and 2, December 2011 and January 2012.

30 EC project GRD1-2001-40449: *Photocatalytic Innovative Coverings Applications for De-Pollution Assessment – PICADA* 2002–2006, www.picada-project.com.

31 Lee J., Mahendra S., Alvarez P.J.J.: Potential environmental and human health impacts of nanomaterials used in construction industry, in *Nanotechnology in Construction NICOM3*, Bartos P.J.M., Bittnar Z., Nemecek J., Smilauer V., Zeman J. (Eds.), Springer-Verlag, Berlin, 2009, pp. 1–11.

32 The International Glassfibre Reinforced Concrete Association: *Specification for the Manufacure, Curing & Testing of Glassfibre Reinforced Concrete (GRC) Products*, iGRCA, Camberley, UK, January 2016.

33 Bartos P.J.M., Zhu W.: Effect of microsilica and acrylic polymer on the ageing of GRC, *Cement and Concrete Composites*, Vol. 18, No. 1, 1996, pp. 31–39.

34 BS EN 1008:2002: Mixing water for concrete. Specification for sampling, testing and assessing the suitability of water, including water recovered from processes in the concrete industry, as mixing water for concrete.

35 Cyr M.F., Peled A., Shah S.P.: Improving performance of glassfibre reinforced extruded composite, *Proceedings of the 12th GRCA Congress*, Dublin, Clarke N. (Ed.), Camberley, 2001, pp. 163–172.

36 Gilbert G.T., Mott J.R.: Filament winding of glassfibre reinforced cement utility poles and other products, in *Fibre-Reinforced Concretes*, de Prisco M., Felicetti R., Plizzari G.A. (Eds.), RILEM Publications s.a.r.l., PRO 039, 2004, pp. 1123–1132.

37 Akkaya Y., Peled A., Shah S.P.: Parameters related to fiber length and process-ing in cementitious composites, *Materials and Structures*, Vol. 33, October 2000, pp. 515–524.

38 Lucem Gmbh: *Lucem Lichtbeton*, Product Brochure, Stolberg, Germany, 2015, www.lucem.de, 2015.

39 Anon.: 10 completely 3D printed houses appear in Shanghai, built in under a day, www.3ders.org, 2014.

40 Hein P.: Thirty years' experience in making rubber moulds for GRC, *Proceedings of the 12th GRCA Congress*, Dublin, 2001, pp. 145–153.

41 Rain C., Kierkegaard P.H.: Adaptive mould: A cost-effective mould system link-ing design and manufacturing of double-curved GFRC panels, *Proceedings of the 16th GRCA Congress*, Dubai, 2015.

42 Bartos P.: *Bond Characteristics in Fibrous Composites with Brittle Matrices*, PhD thesis, University of Southampton, 1976.

43 Bartos P.: Effects of changes in fibre strength and bond characteristics due to age-ing on fracture mechanism of GRC, *Proceedings of the Symposium on Durability of Glass Fiber Reinforced Concrete*, Diamond S. (Ed.), Prestressed Concrete Institute, Chicago, November 1985, pp. 126–146.

44 Bartos P.J.M.: Brittle matrix composites reinforced with bundles of fibres, in *From Materials Science to Construction Materials Engineering*, Maso J.C. (Ed.), September 1987, Chapman and Hall, London, Vol. 3, pp. 539–547.

45 Bartos P.J.M., Duris M.: Inclined tensile strength of steel fibres in a cement-based composite, *Composites*, Vol. 25, No. 10, 1994, pp. 945–952.

46 Duris M.: *Micro-Mechanics of Fracture of Inclined Fibres in a Cement-based Composite*, PhD Thesis, University of Paisley, August 1993.

47 Anon.: GFRC Online, www.gfrconline.com, 2015.

48 Ridd P.: Private communication, 2015.

49 Bartos P.J.M., Zhu W.: Assessment of interfacial microstructure and bond properties in aged GRC using a novel micro-indentation method, *Cement and Concrete Research*, Vol. 27, No. 11, 1997, pp. 1701–1711.

50 Trtik P., Bartos P.J.M., Reeves I.: Use of focused ion beam (FIB) techniques for production of diamond probe for nanotechnology-based single filament push-out tests, *Journal of Materials Science*, Vol. 19, No. 10, May 2000, pp. 903–906.

51 EN 1170 Part 1:1998: Precast concrete products. Test method for glass fibre-reinforced cement. Measuring the consistency of the matrix. 'Slump test' method.

52 Vetrotex Espana SA: *Cem-FIL GRC Technical Data*, December 1998.

53 EN 1170 Part 6:1998: Precast concrete products. Test method for glass fibre-reinforced cement. Determination of the absorption of water by immersion and determination of the dry density.

54 International Glass fibre Concrete Association: *Methods of Testing Glassfibre Reinforced Concrete (GRC) Material*, GRCA, Camberely, UK, April 2013.

55 Anon.: GFRC Online, www.gfrconline.com, 2013.

56 Prestressed Concrete Institute: *Recommended Practice for Glass Fiber Reinforced Concrete Panels*, MNL-128, USA, 4th Edition, 2001.

57 Ridd P.: GRC Noise Barriers, GRCA techNOTE 5, *Concrete*, June 2009, pp. 14–15.

58 EN 1170 Parts 4 and 5: 1998: Precast concrete products. Test method for glassfibre reinforced cement. Measuring bending strength.

59 Jones G.: *Practical Design Guide for GRC using Limit State Theory*. International Glassfibre Reinforced Concrete Association, Camberley, UK, November 2004.

60 EN 1170 Part 8: 2008: Test method for glassfibre reinforced cement. Cyclic weathering type test.

61 EN 13501-1: 2007 Fire classification of construction products and building elements. Classification using data from reaction to fire tests (+A1: 2009).

62 Warrington Fire: *Glass Reinforced Concrete: Classification of Reaction to Fire Performance*, Report by Warrington Fire, International Glass fibre Concrete Association, Camberley, 2008.

63 EN 1170 Part 7: 1998: Test method for glassfibre reinforced cement. Measurement of extremes of dimensional variations due to moisture content.

64 White I: Private communication, 2016.

65 Ferry R., Parrott L.: *Defining and Improving Environmental Performance in the Concrete Industry*, DETR Project 39/3/437, CC1553, UK, 1999.

66 European Commission: *Clean Air Policy Package*, 18 December 2013, http://ec.europa.eu/environment/air/clean_air_policy.htm.

67 Anon.: *Guide to Fixings for Glassfibre Reinforced Cement Cladding*, Glassfibre Reinforced Concrete Association, Wigan, 1994.

68 Curiger P.: *Glassfibre Reinforced Concrete: Practical Design and Structural Analysis*, Beton Verlag for Fachvereinigung Faserbeton e.V., Dusseldorf, 1995.

69 Hanley M.: 3D detailing of GRC elements, GRCA techNOTE, *Concrete*, February 2009, pp. 10–11.

70 BS 8297:2000: Code of practice for Design and installation of non-loadbearing precast concrete cladding.

71 The Health and Safety Executive: *Cement*, Construction Information Sheet No. 26, rev. 2, Sudbury, UK, February 2012.

72 The Health and Safety Executive: *The Control of Substances Hazardous to Health Regulations 2002. Approved Code of Practice and Guidance*, 6th Edition, 2013, UK.

73 Health and Safety Laboratory: *An Inventory of Fibres to Classify Their Potential Hazard and Risk*, Health and Safety Executive Research Report 503, HSE, Sudbury, UK, 2006.

Appendix A

A. CALIBRATION OF GRC SPRAY EQUIPMENT

The strength of GRC composites depends upon the glass fibre content, which for all hand spray operations is generally about 5% by weight of the finished product.

Before starting to spray, it is necessary to calibrate the slurry spray and glass depositor outputs using the bag and bucket tests.

For a typical 12 kg/min slurry output, the glass depositor output should be approximately 630 g/min. In some specifications, the 5% is the *minimum glass content* allowed. It is suggested to use a target glass content of 5.3% in such a case.

A.1 Bag test

This is used to set the correct rate of delivery of chopped fibre from the glass depositor.

A.1.1 Equipment

Balance: Capacity of 1 kg (accurate to within 1 g)
Plastic bag: 600 mm × 1000 mm approximately

A.1.2 Method

The test should be carried out under actual running conditions of the spraying equipment.

1. Weigh an empty bag (W grams).
2. Feed chopped glass into the bag for 15 seconds.
3. Weigh bag and fibre (G grams).

A.1.3 Glass output

Calculate the glass output as $(G - W) \times 4$ g/min. Adjust air pressure to glass depositor until required output is achieved. Note the pressure setting.

A.2 Bucket test

This is used to measure the output of the slurry during the spraying process.

A.2.1 Equipment

Balance: Capacity of not less than 12 kg (accurate to within 50 g or less)
Plastic bucket: 10 litre capacity

A.2.2 Method

The test should be carried out under actual running conditions.

1. Weigh an empty bucket (W grams).
2. Spray slurry into the bucket for 30 seconds.
3. Weigh bucket and slurry (S grams).

A.2.3 Slurry output

Calculate the slurry output as $(S - W) \times 2$ kg/min. Adjust the output of the pump until the required output is achieved. Note the pump setting. Typical settings for fibre content of 5% are shown in Table A.1.

A.2.4 Calculated examples

Glass output

$$\frac{\text{Slurry output } (\text{kg/min}) \times \text{glass content } (\%) \text{ kg/min}}{100 - \text{glass content } (\%)}$$

Example of glass output calculation

If the glass content should be 5% and the slurry output is 12.6 kg/min, the required glass output is

$$\frac{12.6 \times 5}{95} = 0.663 \, \text{kg/min} = 166 \, \text{g per } 15 \text{s}$$

Table A.1 Bag and bucket calibration data or 5% glass content [32]

Glass output		Slurry output	
g/15 s	kg/min	kg/min	kg/30 s
130.0	0.52	9.88	4.940
132.5	0.53	10.07	5.035
135.0	0.54	10.26	5.130
137.5	0.55	10.45	5.225
140.0	0.56	10.64	5.320
142.5	0.57	10.83	5.415
145.0	0.58	11.02	5.510
147.5	0.59	11.21	5.605
150.0	0.60	11.40	5.700
152.5	0.61	11.59	5.795
155.0	0.62	11.78	5.890
158.0[a]	0.632[a]	12.00[a]	6.000[a]
160.0	0.64	12.16	6.080
162.5	0.65	12.35	6.175
165.0	0.66	12.54	6.270
167.5	0.67	12.73	6.365
170.0	0.68	12.92	6.460
172.5	0.69	13.11	6.555
175.0	0.70	13.30	6.650
177.5	0.71	13.49	6.745
180.0	0.72	13.68	6.840
182.5	0.73	13.87	6.935
185.0	0.74	14.06	7.030
187.5	0.75	14.25	7.125
190.0	0.76	14.44	7.220
192.5	0.77	14.63	7.315
195.0	0.78	14.82	7.410
197.5	0.79	15.01	7.505
200.0	0.80	15.20	7.600
202.5	0.81	15.39	7.695
205.0	0.82	15.58	7.790
207.5	0.83	15.77	7.885
210.0	0.84	15.96	7.98

[a] Typical output for concentric spray guns.

Slurry output

$$\frac{\text{Glass output (kg/min)} \times (100 - \text{glass content(\%)}) \,(\text{kg/min})}{\text{Glass content (\%)}}$$

Example of slurry output calculation

If the glass output is 0.7 kg/min and the glass content should be 5%, the required slurry is

$$\frac{0.7 \times 95}{5} = 13.3 \text{ kg/min}$$

Note:

The chopped fibres from the bag test are scrap and are not suitable for premix.
The slurry can be returned to the pump hopper.
The bag and bucket test should be carried out whenever there is a change in
   the mix, whether deliberate or accidental.

A.2.5 Basic procedure

1. Set the glass depositor air pressure gauge to the required level.
2. Carry out the bag test for the glass fibre.
3. Read the required slurry output from Table A.1.
4. Set slurry output using bucket test.

Note: If the glass depositor output falls when using the same air pressure, either the air motor or the filter to the air motor requires maintenance, or the oil bottle feeding the air motor has run dry.

A.3 Mini-slump test: measuring flow of slurry

Scope

The test provides an easy way to check the ability of the fresh GRC to be sprayed. This test is not always suitable for polymer mixes.

The consistency of the mix can affect the spray characteristics and hence the pressures used in atomisation of the slurry.

Maintaining a constant slump means that the ability of the mix to be sprayed easily will be maintained. This will make spraying easier and compaction more efficient.

A.3.1 Equipment

Open-ended Perspex tube: internal Ø 57 mm, external Ø 65 mm, length 55 mm.

Perspex target plate: 30 × 30 cm engraved with a series of concentric circles of diameters 65, 85, 108, 125, 145, 165, 185, 205 and 225 mm, numbered 0–8, respectively.

A.3.2 Method

The plastic tube is placed centrally onto the Plexiglas target plate, held down firmly and filled with slurry. If necessary, air bubbles are expelled by gently tamping the mix. The slurry top should be levelled off using the edge of a spatula.

The tube is lifted vertically off the plate with a slow continuous motion, allowing the slurry to flow over the concentric circles on the target plate.

The slump is measured by the number of rings covered by the slurry. Standard formulations normally give 2–3 rings, but it is up to the manufacturer to achieve consistent results for the mix that they require.

A.3.3 Notes

The consistency and hence the quality of the mix can be affected by

- Cement age and type:
  - Cold cement – low strengths.
  - Warm cement – false sets.
- Sand grading: Use the correct clean and dry grade. Dirty or wet sand can affect workability and strength. A high fines content increases water demand, as does the presence of clay particles.
- Water temperature:
  - Too cold – can retard the setting.
  - Too hot – can cause the mix to develop a 'flash' set.
- Superplasticisers: Match the most suitable superplasticisers with the cement/binder used so as to obtain the best extended slump values.
- Polymer: Store in the conditions recommended by the supplier.
- Mixing time: If an ammeter is attached to the mixer, the power required to mix the formulation can be monitored more closely, and more uniform mixes can be produced.

Any change in consistency during spraying should immediately be reported to the management and checks made on the slurry output and quality.

Note: If a 'false set' occurs in the mixer, stop mixing for 30 seconds and then re-mix for 30 seconds.

Appendix B

B. DETERMINATION OF GLASS CONTENT OF UNCURED GRC

B.1 Scope

This GRCA standard describes a method for the determination of the content of glass fibres (by weight) in a GRC material in the uncured, green state [65].

(An alternative method for measuring the fibre content of fresh GRC and calibrating the equipment is contained in EN 1170 – 3: Measuring the fibre content of sprayed GRC.)

B.2 Definitions

Green state: The stage in the manufacture of GRC when all physical processes which would alter the composition of the material are complete, whilst permitting the fibre to be separated from the matrix by the action of running water. This condition usually lasts up to 2 hours after the completion of manufacture under ambient conditions.

Apparatus: A laboratory balance capable of weighing 1000 grams in increments of 0.1 grams and with an accuracy of ±0.05 grams.

A laboratory oven equipped with forced air circulation and ventilation, capable of achieving a temperature of at least 300°C.

or

A laboratory muffle furnace equipped with ventilation, capable of achieving a temperature of 500 ± 20°C.

Mesh baskets (rectangular, 175 mm long × 100 mm wide × 25 mm deep) made from 3 mm thick stainless steel wire mesh (Figure B.1).

Figure B.1 Stainless steel mesh basket for washing out of the test specimen.

B.3   Test specimen

Preparation

The test specimens shall be taken from the finished product or, where this is not practicable, from a test board prepared so that it represents the product in composition, manufacturing process and thickness. The specimen shall be cut from the product or test panel in green state using a sharp knife or other means resulting in the test specimen having clean, cut edges.

Dimension of a test specimen

The specimen size shall be nominally 150 mm × 50 mm.

Number of test specimens

No fewer than three test specimens shall be taken from a product or test board to obtain a mean value of glass content for that product or test board. The specimens should be chosen so as to be as representative as possible of the total area of the product or test board.

B.4   Test procedure

The test shall be performed immediately after cutting of the specimen from the product or test board.

B.4.1 A dry mesh basket is placed on the balance and its mass (M_2) is recorded.

B.4.2 The test specimen is placed into the basket and weighed and its mass (M_2) is recorded.

B.4.3 The basket with the test specimen is placed under a stream of running water and the GRC is worked with fingers to facilitate its break up. Care must be taken to ensure that no glass is lost.

B.4.4 When all the cement and other solid particles other than the glass fibres have been washed away, the basket and its contents are dried up to constant weight using either a laboratory oven at a temperature above 300°C (approximately 1 hour) or a muffle furnace at a temperature not greater than 520°C (approximately 5 minutes).

B.4.5 On completion of the drying, the basket with its contents is removed from the oven or furnace and allowed to cool to room temperature (preferably in a desiccator).

B.4.6 A visual check is made to confirm that the glass is clean. If necessary, any residual sand is removed by working the glass fibre by hand and shaking the tray.

B.4.7 The dried tray with glass fibres is weighed and its mass (M_3) is recorded.

B.5   Calculation and expression of results

Glass content (by weight)

The glass content (by weight) is calculated using the formula

$$\text{Glass content} = \left(\% \ w \ / \ w\right) = \frac{(M_3 - M_1)}{(M_2 - M_1)} \quad 100$$

where

M_1 = Mass of basket (g)
M_2 = Mass of basket + specimen (g)
M_3 = Mass of basket + dry glass (g)

B.6   Test report

The test report shall include a reference to this standard and the following particulars, as necessary:

1. Product or test board identification mark, date of test and other pertinent data concerning the tested material.
2. The number of test specimens taken as a sample.
3. The arithmetic mean value of the glass content of all the results from the product or board tested and the range of results.

Appendix C

C. DETERMINATION OF FLEXURAL PROPERTIES OF GRC

C.1 Scope

This GRCA standard describes a method for the determination of the limit of proportionality (LOP) and the modulus of rupture (MOR) of GRC material in the form of rectangular elements cut from sheets or moulded directly.

Note: The four-point loading imposes pure bending forces over the middle third of the test specimen and is preferred to the three-point loading test in which the stress is concentrated at the centre.

This procedure relates to the use of a typical universal tensile test machine by which the load deflection curve is recorded automatically.

C.2 Definitions

C.2.1 *Limit of Proportionality*
The applied flexural stress which corresponds to a point where the load/deflection curve begins to deviate from linearity.

C.2.2 *Modulus of Rupture*
The flexural stress developed when load reaches the maximum.

C.3 Apparatus

C.3.1 A standard testing machine properly constructed and calibrated, which can be operated at a constant crosshead rate and in which the error does not exceed ±1% for indicated loads and not exceed 2% for indicated

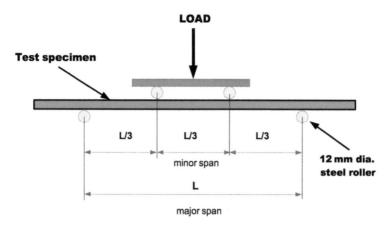

Figure C.1 Standard test for GR in bending – general layout.

deflections. The machine shall comply with the requirements of BS EN ISO 7500 2004.

C.3.2 A bending test jig (as shown schematically in Figure C.1). The supports and loading rollers shall be at least as wide as the test specimen and be designed such that the forces applied to the specimen will be perpendicular to the surface of the specimen and applied without eccentricity. The radius of the loading rollers shall be at least 6 mm. The distance between the supports (L) should be adjustable.

C.4 Test specimen

The test specimens shall be taken from the finished product or where this is not practicable from a test board prepared so that it represents the product in composition, manufacturing process, curing and thickness.

The test specimens shall be cut from the cured finished product or cured test board using a silicon carbide saw or other appropriate equipment with water cooling to avoid affecting the properties of the specimens. The specimens shall be rectangular with parallel sides which are perpendicular with the mould or machine face of the specimen.

C.4.1 *Dimensions of test specimens*: The length of the specimen shall be not less than 25 mm and not more than 50 mm greater than the major span dimension given in Table C.1 for the thickness of the specimen. (Typically, the thickness is 8–10 mm and no greater than 12.5 mm.) The width b should be 50 ± 2 mm.

Table C.1 Major and minor span and crosshead speed for various specimen thicknesses

Nominal specimen thickness (mm)	Major span (mm)	Minor span (mm)	Crosshead speed (mm/min)
Up to 6.7	135	45.0	1.5–3.0
6.8–10.0	200	66.7	1.5–3.0
10.1–12.5	250	83.3	1.5–3.0
12.6–15.0	300	100.0	3.0–5.0
15.1–17.5	350	116.7	3.0–5.0
17.6–20.0	400	133.3	3.0–5.0

C.4.2 *Anisotropy of the material* If the GRC tested exhibits a visible anisotropy within the plane of the test board/sheet, and the direction in which the greatest strength lies is known from the experience of the manufacturer or the production process, then at least three test specimens shall be taken whose length is parallel to that direction, and three test specimens shall also be taken whose length is normal to that direction. The direction of the greatest strength should be marked on the sheet before testing.

C.5   Number of test specimens

At least four test specimens should be used. Two of these should be tested with the mould or machine face of the specimen in contact with the major span rollers, and two should be tested with the minor span rollers.

C.6   Procedure

C.6.1 Conditioning of test specimens

The test specimens shall be conditioned by soaking in water at room temperature for a period of between 4 and 24 hours and tested wet. The test shall be performed within 5 minutes of removing the specimens from the soaking procedure. Removing the surface water with a towel is permitted.

An exception to the conditioning in water can be made in the case when the GRC mix contains a polymer for self-curing and the test is used primarily for quality control during production. Wet conditioning should be used whenever comparisons between different GRC mix designs or production processes are made.

C.6.2 Testing procedure

- Set the major and minor spans of the test jig to correspond with Figure C.1. The loading rollers and supports should be aligned so that the axes of the cylindrical surface are parallel.

- Place the test specimen symmetrically across the two parallel supports, ensuring that the length of the test specimen is at right angles to each of these and that equal lengths of the specimens project outside of the rollers.

- Set the crosshead speed of the machine to that indicated in Table C.1.

- Apply the load at constant crosshead speed to failure, continuously recording the load-deflection curve. The load range should be chosen so that the LOP load occurs at not less than 30% of the full-scale load range.

- Record the load (W_1) at which the load-deflection curve deviates from linearity (LOP load) and also the maximum load (W_2) obtained (MOR load). This is done automatically on most current testing machines.

- Separate the failed test piece and measure the specimen thickness at the failure zone in three places to the nearest 0.1 mm. Calculate and record the average. Measure the specimen width to the nearest 0.1 mm. These measurements should be taken at or near the failure location, taking care not to choose places where the specimen may have expanded during the test.

C.7 Calculation and expression of results[1]

C.7.1 Limit of proportionality

The limit of proportionality (MPa) is calculated using the equation

$$2LOP = W_1 L / bd$$

where
W_1 = LOP load (N), i.e., the load at which the load-deflection curve begins to deviate from linearity
L = major span (mm)
b = width (mm)
d = thickness (mm)

C.7.2 Modulus of rupture

The MOR (MPa) is calculated using the equation

$$MOR = W_2 L / bd^2$$

where
W_2 = MOR load (N)
L = major span (mm)
b = width (mm)
d = thickness (mm)

C.7.3 Directionality ratio

Where anisotropic material is tested, the ratio of the mean value for both LOP and MOR from each direction shall be given as the directionality ratio.

C.8 Test Report

The test report shall include a reference to this GRCA standard. The report shall also refer to the following particulars, as necessary.

1. Product or test board identification mark, date of test and other pertinent data concerning the tested material.
2. The number of test specimens taken as a sample.
3. The direction from which the specimens were taken, if anisotropic material is tested.
4. The individual values of the LOP and MOR for each specimen tested.

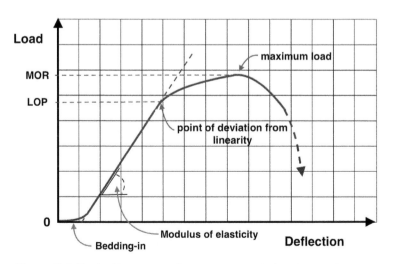

Figure C.2 Typical load-extension curve recorded in the test for bending strength.

5. The minimum and arithmetic mean value of the LOP and MOR with one face of the specimens in contact with the major span supports.

6. The minimum and arithmetic mean value of the LOP and MOR with the reverse faces of the specimens in contact with the major span supports.

7. The minimum and overall arithmetic mean of LOP and MOR calculated from all the specimens tested.

Note

1   Purpose-built testing equipment can be purchased with built-in software that will automatically calculate the LOP and MOR plus strain to LOP, strain to MOR and Young's modulus.

Appendix D

D. DETERMINATION OF THE DRY AND WET BULK DENSITY, WATER ABSORPTION AND APPARENT POROSITY OF GRC

D.1 Scope

A single method for the determination of dry and wet bulk density, water absorption and apparent porosity of GRC.

D.2 Apparatus

D.2.1 A laboratory balance capable of weighing 1000 g in increments of 0.1 g and with an accuracy of ±0.05 g. The balance must be capable of weighing a test specimen suspended in water.

D.2.2 A suitable holder for suspending the test specimen in water.

D.2.3 A laboratory oven, equipped with forced air circulation, capable of achieving a temperature of 110 ± 5°C.

D.2.4 A desiccator capable of holding several 100 × 100 mm test pieces.

D.3 Test specimen

The test specimen shall be taken from the finished product or, where this is not practicable, from a test board prepared so that it represents the product in its composition, manufacturing process, curing and thickness. The test specimens shall be cut out from the cured finished product or cured test board using a silicon carbide saw or other appropriate equipment.

D.3.1 *Dimensions of a test specimen*: The specimen size shall be not less than 50 × 50 mm (nominal sizes) and preferably 100 × 100 mm.

D.3.2 *Number of test specimens*: No fewer than two test specimens shall be taken from a product or test board to obtain mean values for the product or the test board. The test specimens should not be taken from adjacent areas. The specimen should be free from visible cracks, fissures or broken edges.

D.4   Test procedure

1. Immerse the specimen in freshwater at a temperature of approximately 20°C until a constant weight is achieved. (This typically takes 7 days.)
2. Determine the mass M1 of the specimen suspended in water. Take the specimen out of the water, quickly remove the surface water with a paper towel and immediately weigh the specimen in air to obtain mass M2.
3. Heat the specimen in an oven at a temperature of 110 ± 5°C to constant weight (1–7 days). Remove the specimen from the oven, allow to cool in a desiccator to room temperature and weigh to obtain mass M3.

D.5   Calculation and expression of results

The properties are calculated using for the following formulae.

D.5.1 Dry bulk density (kg/m³) = $\dfrac{M_3 \times 1000}{M_2 - M_1}$

D.5.2 Wet bulk density (kg/m³) = $\dfrac{M_2 \times 1000}{M_2 - M_1}$

D.5.3 Water absorption (% by weight) = $\dfrac{M_2 - M_3 \times 100}{M_3}$

D.5.4 Apparent porosity (% by volume) = $\dfrac{(M_2 - M_3) \times 100}{M_2 - M_1}$

D.6   Test Report

The test report shall include a reference to this standard and the following particulars, as necessary.

1. Product or test board identification mark, date of test and other pertinent data concerning the tested material.
2. The number of test specimens taken as a sample.
3. The arithmetic mean value of the water/solid ratio of all the results from the product or board tested and the range of results.

GLASSFIBRE REINFORCED CONCRETE

A PROVEN PAST...

...AND AN EXCITING FUTURE

Since 1978 the International Glassfibre Reinforced Concrete Association has been there to support manufacturers and all those with an interest in GRC. We have developed quality and testing standards recognised throughout the world and provide reassurance to specifiers and users of the material through our "Full Member" assessment scheme. As GRC becomes ever more popular we will continue to lead in the innovation and development of this unique composite.

GRCA
INTERNATIONAL

The International Glassfibre Reinforced Concrete Association, PO Box 1454, Northampton, NN2 1DZ. United Kingdom

Index